Hawking Radiation

3

Roman Andie

Lighthouse Books for
Translation and Publishing

LIGHTHOUSE
FOR TRANSLATISION AND PUBLISHING
للترجمة والنشر

Hawking Radiation 3 by Roman Andie

Email: info@lighthousebooks.org

ISBN 13: 9783592132243

ISBN 10: 3592132245

Time reversed cognition

Time reflection yields time reversed and spatially reflected sensory-cognitive representations. When mental image dies it is replaced with its time-reversal at opposite boundary of its CD. The observation of these representations could serve as a test of the theory.

There is indeed some evidence for this rather weird looking time and spatially reversed cognition.

1. I have a personal experience supporting the idea about time reversed cognition. During the last psychotic episodes of my "great experience" I was fighting to establish the normal direction of the experienced time flow. Could this mean that for some sub-CDs the standard arrow of time had reversed as some very high level mental images representing bodily me died and was re-incarnated?
2. The passive boundary of CD corresponds to static observing self - kind of background - and active boundary the dynamical - kind of figure. Figure-background division of mental image in this sense would change as sub-self dies and re-incarnates since figure and background change their roles. Figure-background illusion could be understood in this manner.
3. The occurrence of mirror writing is well known phemonenon (my younger daughter was reverse writer when she was young). Spatial reflections of MEs are also possible and might be involved with mirror writing. The time reversal would change the direction of writing from right to left.
4. Reverse speech would be also a possible form of reversed cognition. Time reversed speech has the same power spectrum as ordinary speech and the fact that it sounds usually gibberish means that phase information is crucial for storing the meaning of speech. Therefore the hypothesis is testable.

Reverse speech

Interestingly, the Australian David Oates claims that so called reverse speech is a real phenomenon, and he has developed entire technology and therapy (and business) around this phenomenon. What is frustrating that it seems impossible to find comments of professional linguistics or neuro-scientits about the claims of Oates. I managed only to find comments by a person calling himself a skeptic believer but it became clear that the comments of this highly rhetoric and highly arrogant commentator did not contain any information. This skeptic even taught poor Mr. Oates in an aggressive tone that serious scientists are not so naive that they would even consider the possibility of taking seriously what some Mr. Oates is saying.

The development of Science can often depend on ridiculously small things: in this case one should find a shielded place (no ridiculing skeptics around) to wind tape recorder backwards and spend few weeks or months to learn to recognize reverse speech if it really is there! Also computerized pattern recognition could be used to make speech recognition attempts objective since it is a well-known fact that brain does feature recognition by completing the data into something which is familiar.

The basic claims of Oates are following.

1. Reverse speech contains temporal mirror images of ordinary words and even metaphorical statements, that these words can be also identified from Fourier spectrum, that brain responds in unconscious manner to these words and that this response can be detected in EEG. Oates classifies these worlds to several categories. These claims could be tested and pity that no professional linguist nor neuroscientist (as suggested by web search) has not seen the trouble of finding whether the basic claims of Oates are correct or not.
2. Reverse speech is complementary communication mode to ordinary speech and gives rise to a unconscious (to us) communication mechanism making lying very difficult. If person consciously lies, the honest alter ego can tell the truth to a sub-self understanding the reverse speech. Reverse speech relies on metaphors and Oates claims that there is general vocabulary. Could this taken to suggest that reverse speech is communication of right brain whereas left brain uses ordinary speech? The notion of semitrance used to model bicameral mind suggests that reverse speech could be communication of higher levels of self hierarchy dispersed inside the ordinary speech. There are also other claims relating the therapy using reverse speech, which sound rather far-fetched but one should not confuse these claims to those which are directly testable.

Physically-reverse speech could correspond to phase conjugate sound waves which together with their electromagnetic counterparts can be produced in laboratory. Phase conjugate waves have rather weird properties due the fact that second law applies in a reversed direction of geometric time. For this reason phase conjugate waves are applied in error correction. ZEO predicts this phenomenon.

Negative energy topological light rays are in a fundamental role in the TGD-based model for Living matter and brain. The basic mechanism of intentional action would rely on time mirror mechanism utilizing the TGD counterparts of phase conjugate waves producing also the nerve pulse patterns generating ordinary speech. If the language regions of brain contain regions in which the the arrow of psychological time is not always the standard one, they would induce phase conjugates of the sound wave patterns associated with the ordinary speech and thus reverse speech.

ZEO-based quantum measurement theory, which is behind the recent form of TGD-inspired theory of Consciousness provides a rigorous basis for this picture. Negative energy signals can be assigned with sub-CDs representing selves with non-standard direction of geometric time and every time when mental image dies, a mental images with opposite arrow of time is generated. It would be not surprising if the reverse speech would be associated with these time reversed mental images.

Figure-background rivalry and time reversed mental images

The classical demonstration of figure-background rivalry is is a pattern experienced either as a vase or two opposite faces. This phenomenon is not the same thing as bi-ocular rivalry in which the percepts associated with left and right eyes produced by different sensory inputs are rivalling. There is also an illusion in which one perceices the dancer to make a pirouette in either counter-clockwise or clockwise direction althought the figure is static. The direction of pirouette can change. In this case time-reversal would naturally change the direction of rotation.

Figure-background rivalry gives a direct support for the TGD-based of self relying on ZEO if the following argument is accepted.

1. In ZEO the state function reduction to the opposite boundary of CD means the death of the sensory mental image and birth of new one, possibly the rivalling mental image. During the sequence of state function reductions to the passive boundary of CD defining the mental image a boundary quantum superposition of rivalling mental images associated with the active boundary of CD is generated.

   In the state function reduction to the opposite boundary the previous mental image dies and is replaced with new one. In the case of bin-ocular rivalry this might be the either of the sensory mental images generated by the sensory inputs to eyes. This might happen also now but also different interpretation is possible.

2. The basic questions concern the time reversed mental image. Does the subject person as a higher level self experience also the time reversed sensory mental image as sensory mental image as one might expect. If so, how the time reversed mental image differs from the mental image? Passive boundary of CD defines quite generally the background - the static observer - and active boundary the figure so that their roles should change in the reduction to the opposite boundary.In sensory

rivalry situation this happens at least in the example considered (vase and two faces).

I have also identified motor action as time reversal of sensory percept. What this identification could mean in the case of sensory percepts? Could sensory and motor be interpreted as an exchange of experiencer (or sub-self) and environment as figure and background?

If this interpretation is correct, figure-background rivalry would tell something very important about consciousness and would also support ZEO. Time reversal would permute figure and background. This might happen at very abstract level. Even subjective-objective duality and first - and third person aspects of conscious experience might relate to the time reversal of mental images. In near death experiences person sees himself as an outsider: could this be interpreted as the change of the roles of figure and background indentified as first and third person perspectives? Could the first moments of the next life be seeing the world from the third person perspective?

An interesting question is whether right- and left hemispheres tend to have opposite directions of Geometric-Time. This would make possible metabolic energy transfer between them making possible kind of flip-flop mechanism. The time-reversed hemisphere would receive negative energy serving as metabolic energy resource for it and the hemisphere sending negative energy would get in this manner positive metabolic energy. Deeper interpretation would be in terms of periodic transfer of negentropic entanglement. This would also mean that hemispheres would provide two views about the world in which figure and background would be permuted.

See the article Time Reversed Self. For a summary of earlier postings see Links to the latest progress in TGD.

posted by Matti Pitkanen @ 12:25 AM

2 Comments:

At 6:02 PM, L. Edgar Otto said...

> I have thought a long time about this division of the human mind, and think that at times one side can teach the other... potentially, along these lines. I think in general the side process differently as if one side applies ideas of quantum physics and the other side more toward the relativistic regime. But this is simplistic for the complexity of the question you raise. What for example is folded inward or outward. Anyway, I wish you were more into the

discussion lately as in facebook with especially the debate on Sabine's blog. And it is clear that new things have come from the microtubil ideas.

At 7:31 PM, Matpitka@luukku.com said...

I am surprised that this division might be there. I only recently realised that this is one of the predictions of ZEO. Interpretation is of course difficult. There are many dualities which might relate to this division: for instance, first and third person aspects of consciousness are analogous to figure background. And right brain might be time reverse of left. This would allow two different views with figure and background exchanged. Very useful.

I like the discussion in Sabine's blog. Especially Sabine' critical view and short comment of Arun, which crystallised what I would have said.

What went wrong with superstrings and theoretical physics is a question which should be discussed thoroughly although it is certainly not a pleasant topic of discussion for fans of superstrings.

The four decades 1975-2015 is very interesting historically since the normal evolution of theoretical physics stopped. What caused the stagnation? Was it the emergence of media technology making possible manipulation and hyping in unforeseen manner one of the key factors. I have also the feeling that theoretical done by relatively small top elite transformed to the analog of popular music in which mediocrits dominate and success is determined by marketing efforts rather than musical content? Nowadays marketing is essential part of visible science. Those who are not working with fashionable topics remain in shadow.

05/19/2015 -

Voevoedski's univalent foundations of mathematics

I found a very nice article about the work of mathematician Voevoedski: he talks about univalent foundations of mathematics. To put in nutshell: the world deals with mathematical proofs and the deep idea is that proofs are like paths in some abstract space. One deals with paths also in homotopy theory. What is remarkable that Voevoedski's work leads to computer programs allowing to check that proof does not contain an error. Something very badly needed when the proofs are getting increasingly complex. I dare to guess that the article is understandable by laymen too.

Russell's problem

The article tells about type theory originated already by Russell, who was led to his paradox with the set consisting of sets which do not contain itself as an element. The fatal question was "Does this set contain itself?"

Russell proposed a solution of the problems by introducing a hierarchy of types. Sets are at the bottom and defined so that they do not contain as set collection of sets. In usual applications of set theory - say in manifold theory - this is always true. Type hierarchy guarantees that you do not put apples and oranges in the same basket.

Voevoedski's idea

Voevoedski's work relates to proof theory and to formalizing what mathematical proof does mean.

Consider demonstration that two sets A and B are equivalent. This means simple thing: construct a one-to-one map between them. Usually one is only interested in the existence of this map but one can also get interested on all manners to perform this map. All manners to make this map define in rather abstract sense a collection of paths between A and B regarded as objects. Single path consists of a collection of the arrows connecting element in A with element in B.

More generally, in mathematical proof theory all proofs of theorem define this kind of collection. In topology all paths connecting two points defined this kind of collection. In this framework Goedel's theorem becomes obvious: given axioms define rules for constructing paths and cannot give the paths connecting arbitrarily chosen two truths.

One can again abstract this process. Just as one can make statements about statements about..., one can build paths between paths, and paths between paths between paths.... Voevoedsky studied this problem in his attempts to formalise mathematics and with a practical goal to develop eventually tools for checking by computer that mathematical proofs are correct.

A more rigorous manner to define path is in terms of groupoid. It is more general notion that that of group since the inverse of groupoid element need not exist. For intense open paths between two points form a group but only closed paths can be given a structure of group.

Voevoedski introduced the notion of infinite groupoid containing paths, paths between paths, ... ad infinitum. Voevoedski talks about univalent foundations. The great idea is that homotopy theory becomes a foundation of mathematics: proofs are paths in some abstract structure. This suggests in my non-professional mind that one can talk about continuous deformations of proofs and that one can classify them to homotopy types with proofs of same homotopy type deformable to each other.

How this relates to Quantum-TGD

What made me fascinated was how closely the basic hierarchies of Quantum-TGD relate to the objects studies in type theory and Voevoedski's approach.

Let us start with set theory and type theory. TGD provides a non-trivial example about types which by the way distinguishes TGD from superstring models. Imbedding space is set, space-time surface are its subsets. "World of Classical Worlds" (WCW) is the key abstraction distinguishing TGD from superstring models where one still tries to deal by working at space-time level.

What has surprised me against and again that super string modellers have spend decades in the landscape instead of making super string models a real theory by introducing loop space as a key notion although it has very nice mathematics: just the existence of Kähler geometry fixes it uniquely: this observation actually led to the realisation that quantum TGD might be unique just from its mathematical existence.

The points of WCW are 3-surfaces and its sets are collections of 3-surfaces. They are of higher type than the sets of imbedding space. There would be no sense in putting points of WCW and of imbedding space in the same basket. But in the set theory before Russell you could in principle do this. We have got as as birth gift the ability to not put cows and tooth brushes into same set. But the ability to take seriously the existence of more abstract types does not seem to be a birth gift.

Voevoedski and others deal with statements about statements about statements…. What is amusing that this vision has direct counterparts in TGD based quantum physics where various hierarchies have taken key role. Some deep ideas seem to burst out simultaneously in totally different contexts! Voevoedski noticed the same thing in his work but with within the realm of mathematics.

Just a list of examples should be enough. Consider first type theory.

1. The hierarchy of infinite primes (integers, rationals) was the first mathematical discovery inspired by TGD inspired theory of consciousness. Infinite primes are constructed by a process analogous to a repeated second quantisation of arithmetic quantum field theory having interpretation as making statements about statements about….. up to arbitrary high order. Hierarchy of space-time sheets of many-sheeted space-time is the classical counterpart. Physics prediction is that higher level of quantisation are part of generalised quantum physics and allow quantum description of macroscopic and even astrophysical objects. The map of the sheets of many-sheeted space-time to single region of Minkowski space defines the contraction of TGD to GRT and is approximate operation: it maps a hierarchy of types to single type and is a violent procedure meaning a loss of information.

    Infinite integers could provide a generalisation of Goedel number numbering in a quantum mathematics based on the replacement of axiomatics with anti-axiomatics: specify what you cannot do instead of what you can do! I wrote about this in earlier posting.

    Infinite rationals of unit real norm lead also to a generalisation of the real number. Each real point becomes infinite dimensional space consisting of all infinite rationals of unit norm but well-defined number theoretic anatomy.

2. There are several very closely related infinite hierarchies. Fractal hierarchy of quantum criticalities (ball at the top of a hill at the top…) and isomorphic super-symplectic sub-algebras with conformal structure. There is infinite fractal hierarchy of conformal gauge symmetry breakings. This defines infinite hierarchy of dark matters with Planck constant $h_{eff} = n \times h$. The algebraic extensions of rationals giving rise to evolutionary hierarchy for physics and perhaps explaining biological evolution define a hierarchy. The inclusions of hyper-finite factors realizing finite

measurement resolution define a hierarchy. Hierarchy of infinite integers and rationals relates also closely to these hierarchies.

3. In TGD-inspired theory of Conscioussness hierarchy of selves having sub-selves (experienced as mental images) having.... This hierarchy relates also very closely to the above hierarchies.

The notion of mathematical operation sequences as path is second key idea in Voevoedski's work. The idea about paths representing mathematical computations, proofs, etc.. is realised quite concretely in TGD quantum physics. Scattering amplitudes are identified as representations for sequences of algebraic operations of Yangian leading from an initial collection of elements of super-symplectic Yangian (physical states) to a final one. The duality symmetry of old fashioned string models generalises to a statement that any two sequences connecting same collections are equivalent and correspond to same amplitudes. This means extremely powerful predictions and it seems that in twistor programs analogous results are obtained too: very many twistor Grassmann diagrams represent the same scattering amplitude.

posted by Matti Pitkanen @ 8:49 PM

05/19/2015 - http://matpitka.blogspot.com/2015/05/more-about-physical-interpretation-of.html#comments

More about physical interpretation of algebraic extensions of rationals

The number theoretic vision has begun to show its power. The basic hierarchies of Quantum-TGD would reduce to a hierarchy of algebraic extensions of rationals and the parameters such as the degrees of the irreducible polynomials characterizing the extension and the set of ramified primes - would characterize quantum criticality and the physics of dark matter as large h_{eff} phases. The identification of preferred p-adic primes as ramified primes of the extension and generalization of p-adic length scale hypothesis as prediction of NMP are basic victories of this vision (see this and this).

By strong form of holography the parameters characterizing string world sheets and partonic 2-surfaces serve as WCW coordinates. By various conformal invariances, one expects that the parameters correspond to conformal moduli, which means a huge simplification of Quantum-TGD since the mathematical apparatus of superstring theories becomes available and number theoretical vision can be realized. Scattering amplitudes can be constructed for a given algebraic extension and continued to various number fields by continuing the parameters which are conformal moduli and group invariants characterizing incoming particles.

There are many un-answered and even un-asked questions.

1. How the new degrees of freedom assigned to the n-fold covering defined by the space-time surface pop up in the number theoretic picture? How the connection with preferred primes emerges?
2. What are the precise physical correlates of the parameters characterizing the algebraic extension of rationals? Note that the most important extension parameters are the degree of the defining polynomial and ramified primes.

1. Some basic notions

Some basic facts about extensions are in order. I emphasize that I am not a specialist.

1.1. Basic facts

The algebraic extensions of rationals are determined by roots of polynomials. Polynomials be decomposed to products of irreducible polynomials, which by definition do not contain factors which are polynomials with rational coefficients. These polynomials are characterized by their degree n, which is the most important parameter characterizing the algebraic extension.

One can assign to the extension primes and integers - or more precisely, prime and integer ideals. Integer ideals correspond to roots of monic polynomials $P_n(x) = x^n + .. a_0$ in the extension with integer coefficients. Clearly, for n=0 (trivial extension) one obtains ordinary integers. Primes as such are not a useful concept since roots of unity are possible and primes which differ by a multiplication by a root of unity are equivalent. It is better to speak about prime ideals rather than primes.

Rational prime p can be decomposed to product of powers of primes of extension and if some power is higher than one, the prime is said to be _ramified_ and the exponent is called ramification index. _Eisenstein's criterion_ states that any polynomial $P_n(x) = a_n x^n + a_{n-1} x^{n-1} + ... a_1 x + a_0$ for which the coefficients a_i, i<n are divisible by p and a_0 is not divisible by p^2 allows p as a maximally ramified prime. mThe corresponding prime ideal is n:th power of the prime ideal of the extensions (roughly n:th root of p). This allows to construct endless variety of algebraic extensions having given primes as ramified primes.

Ramification is analogous to criticality. When the gradient potential function V(x) depending on parameters has multiple roots, the potential function becomes proportional a higher power of x-x_0. The appearance of power is analogous to appearance of higher power of prime of extension in ramification. This gives rise to cusp catastrophe. In fact, ramification is expected to be number theoretical correlate for the quantum criticality in TGD framework. What this precisely means at the level of space-time surfaces, is the question.

1.2 Galois group as symmetry group of algebraic physics

I have proposed a long time ago that Galois group acts as fundamental symmetry group of quantum TGD and even made clumsy attempt to make this idea more precise in terms of the notion of number theoretic braid. It seems that this notion is too primitive: the action of Galois group must be realized at more abstract level and WCW provides this level.

First some facts (I am not a number theory professional, as the professional reader might have already noticed!).

1. Galois group acting as automorphisms of the field extension (mapping products to products and sums to sums and preserves norm) characterizes the extension and its elements have maximal order equal to n by algebraic n-dimensionality. For instance, for complex numbers Galois group acs as complex conjugation. Galois group has natural action on prime ideals of extension mapping them to each other and preserving the norm determined by the determinant of the linear map defined by the multiplication with the prime of extension. For instance, for the quadratic extension $Q(5^{1/2})$ the norm is $N(x+5^{1/2}y)=x^2-5y^2$: not that number theory leads to Minkowkian metric signatures naturally. Prime ideals combine to form orbits of Galois group.
2. Since Galois group leaves the rational prime p invariant, the action must permute the primes of extension in the product representation of p. For ramified primes the points of the orbit of ideal degenerate to single ideal. This means that primes and quite generally, the numbers of extension, define orbits of the Galois group.

Galois group acts in the space of integers or prime ideals of the algebraic extension of rationals and it is also physically attractive to consider the orbits defined by ideals as preferred geometric structures. If the numbers of the extension serve as parameters characterizing string world sheets and partonic 2-surfaces, then the ideals would naturally define subsets of the parameter space in which Galois group would act.

The action of Galois group would leave the space-time surface invariant if the sheets co-incide at ends but permute the sheets. Of course, the space-time sheets permuted by Galois group need not co-incide at ends. In this case the action need not be gauge action and one could have non-trivial representations of the Galois group. In Langlands correspondence these representation relate to the representations of Lie group and something similar might take place in TGD as I have indeed proposed.

Remark: Strong form of holography supports also the vision about quaternionic generalization of conformal invariance implying that the adelic space-time surface can be constructed from the data associated with functions of two complex variables, which in turn reduce to functions of single variable.

If this picture is correct, it is possible to talk about quantum amplitudes in the space defined by the numbers of extension and restrict the consideration to prime ideals or more general integer ideals.

1. These number theoretical wave functions are physical if the parameters characterizing the 2-surface belong to this space. One could have purely number theoretical quantal degrees of freedom assignable to the hierarchy of algebraic extensions and these discrete degrees of freedom could be fundamental for Living matter and understanding of Consciousness.
2. The simplest assumption that Galois group acts as a gauge group when the ends of sheets co-incide at boundaries of CD seems however to destroy hopes about non-trivial number theoretical physics but this need not be the case. Physical intuition suggests that ramification somehow saves the situation and that the non-trivial number theoretic physics could be associated with ramified primes assumed to define preferred p-adic primes.

2. How new degrees of freedom emerge for ramified primes?

How the new discrete degrees of freedom appear for ramified primes?

1. The space-time surfaces defining singular coverings are n-sheeted in the interior. At the ends of the space-time surface at boundaries of CD however the ends co-incide. This looks very much like a critical phenomenon.

    Hence the idea would be that the end collapse can occur only for the ramified prime ideals of the parameter space - ramification is also a critical phenomenon - and means that some of the sheets or all of them co-incide. Thus the sheets would co-incide at ends only for the preferred p-adic primes and give rise to the singular covering and large h_{eff}. End-collapse would be the essence of criticality! This would occur, when the parameters defining the 2-surfaces are in a ramified prime ideal.

2. Even for the ramified primes there would be n distinct space-time sheets, which are regarded as physically distinct. This would support the view that besides the space-like 3-surfaces at the ends the full 3-surface must include also the light-like portions connecting them so that one obtains a closed 3-surface. The conformal gauge

equivalence classes of the light-like portions would give rise to additional degrees of freedom. In space-time interior and for string world sheets they would become visible.

For ramified primes n distint 3-surfaces would collapse to single one but the n discrete degrees of freedom would be present and particle would obtain them. I have indeed proposed number theoretical second quantization assigning fermionic Clifford algebra to the sheets with n oscillator operators. Note that this option does not require Galois group to act as gauge group in the general case. This number theoretical second quantization might relate to the realization of Boolean algebra suggested by weak form of NMP (see this).

3. About the physical interpretation of the parameters characterizing algebraic extension of rationals in TGD framework

It seems that Galois group is naturally associated with the hierarchy $h_{eff}/h=n$ of effective Planck constants defined by the hierarchy of quantum criticalities. n would naturally define the maximal order for the element of Galois group. The analog of singular covering with that of $z^{1/n}$ would suggest that Galois group is very closely related to the conformal symmetries and its action induces permutations of the sheets of the covering of space-time surface.

Without any additional assumptions the values of n and ramified primes are completely independent so that the conjecture that the magnetic flux tube connecting the wormhole contacts associated with elementary particles would not correspond to very large n having the p-adic prime p characterizing particle as factor ($p=M_{127}=2^{127}-1$ for electron). This would not induce any catastrophic changes.

TGD based physics could however change the situation and reduce number theoretical degrees of freedom: the intuitive hypothesis that p divides n might hold true after all.

1. The strong form of GCI implies strong form of holography. One implication is that the WCW Kähler metric can be expressed either in terms of Kähler function or as anti-commutators of super-symplectic Noether super-charges defining WCW gamma matrices. This realizes what can be seen as an analog of Ads/CFT correspondence. This duality is much more general. The following argument supports this view.
    1. Since fermions are localized at string world sheets having ends at partonic 2-surfaces, one expects that also Kähler action can be expressed as an effective stringy action. It is natural to assume that string area action is replaced with the area defined by the effective metric of string world sheet expressible as anti-commutators of Kähler-Dirac gamma matrices defined by contractions of canonical momentum currents with imbedding space gamma matrices. It string tension is proportional to h_{eff}^2, string length scales as h_{eff}.

2. AdS/CFT analogy inspires the view that strings connecting partonic 2-surfaces serve as correlates for the formation of - at least gravitational - bound states. The distances between string ends would be of the order of Planck length in string models and one can argue that gravitational bound states are not possible in string models and this is the basic reason why one has ended to landscape and multiverse non-sense.

2. In order to obtain reasonable sizes for astrophysical objects (that is sizes larger than Schwartschild radius $r_s=2GM$) For $h_{eff}=h_{gr}=GMm/v_0$ one obtains reasonable sizes for astrophysical objects. Gravitation would mean quantum coherence in astrophysical length scales.

3. In elementary particle length scales the value of h_{eff} must be such that the geometric size of elementary particle identified as the Minkowski distance between the wormhole contacts defining the length of the magnetic flux tube is of order Compton length - that is p-adic length scale proportional to $p^{1/2}$. Note that dark physics would be an essential element already at elementary particle level if one accepts this picture also in elementary particle mass scales. This requires more precise specification of what darkness in TGD sense really means.

One must however distinguish between two options.

1. If one assumes $n \approx p^{1/2}$, one obtains a large contribution to classical string energy as $\Delta \sim m_{CP2}^2 L_p / \hbar^2_{eff} \sim m_{CP2}/p^{1/2}$, which is of order particle mass. Dark mass of this size looks un-feasible since p-adic mass calculations assign the mass with the ends wormhole contacts. One must be however very cautious since the interpretations can change.

2. Second option allows to understand why the minimal size scale associated with CD characterizing particle correspond to secondary p-adic length scale. The idea is that the string can be thought of as being obtained by a random walk so that the distance between its ends is proportional to the square root of the actual length of the string in the induced metric. This would give that the actual length of string is proportional to p and n is also proportional to p and defines minimal size scale of the CD associated with the particle. The dark contribution to the particle mass would be $\Delta m \sim m_{CP2}^2 L_p / \hbar^2_{eff} \sim m_{CP2}/p$, and completely negligible suggesting that it is not easy to make the dark side of elementary visible.

4. If the latter interpretation is correct, elementary particles would have huge number of hidden degrees of freedom assignable to their CDs. For instance, electron would have $p=n=2^{127}-1 \approx 10^{38}$ hidden discrete degrees of freedom and would be rather intelligent system - 127 bits is the estimate- and thus far from a point-like idiot of standard physics. Is it a mere accident that the secondary p-adic time scale of electron is .1 seconds - the fundamental biorhythm - and the size scale of the minimal CD is slightly large than the circumference of Earth?

Note however, that the conservation option assuming that the magnetic flux tubes connecting the wormhole contacts representing elementary particle are in $h_{eff}/h=1$ phase can be considered as conservative option.

See the article More about physical interpretation of algebraic extensions of rationals. For a summary of earlier postings see Links to the latest progress in TGD.

posted by Matti Pitkanen @ 1:53 AM

05/14/2015 - http://matpitka.blogspot.com/2015/05/quantum-mathematics-in-tgd-universe.html#comments

Quantum Mathematics in the TGD Universe

Some comments about quantum mathematics, quantum Boolean thinking and computation as they might happen at fundamental level.

1. One should understand how Boolean statements $A \to B$ are represented. Or more generally: How a computation like procedure leading from a collection A of math objects collection B of math objects takes place? Recall that in computations the objects coming in and out are bit sequences. Now one have computation like process. $\to$ is expected to correspond to the arrow of time.

   If fermionic oscillator operators generate Boolean basis, zero energy ontology is necessary to realize rules as rules connecting statements realized as bit sequences. Positive energy ontology would allow only statements A,B but not statements $A \to B$ about them. ZEO allows also to avoid restrictions due to fermion number conservation and its well-definedness.

   Collection A is at the passive boundary of CD and not changed in state function reduction sequence defining self and B is at the active one. As a matter fact, it is not single statement but a quantum superpositions of statements B, which resides there! In the quantum jump selecting single B at the active boundary, A is replaced with a superposition of A:s: self dies and re-incarnates as more negentropic entity. Q-computation halts.

   That both a and b cannot be known precisely is a quantal limitation to what can be known: philosopher would talk about epistemology here. The different pairs (a,b) in superposition over b:s are analogous to different implications of a. Thinker is doomed to always live in a quantum cognitive dust and never be quite sure of.

2. What is the computer like structure now? Turing computer is discretized 1-D time-like line. This quantum computer is superposition of 4-D space-time surfaces with the basic computational operations located along it as partonic 2-surfaces defining the algebraic operations and connected by fermion lines representing signals. Also string world sheets are involved. In some key aspects this is very similar to ordinary computer. By strong form of holography computations use only data at string world sheets and partonic 2-surfaces.

3. What is the computation? It is sequence of repeated state function reduction leaving the passive boundary of CD intact but affecting the position (moduli) of upper boundary of CD and also the parts of zero energy states there. It is a sequence of unitary processes delocalizing the active boundary of CD followed by localization but no reduction. This the counterpart for a sequence of reductions leaving quantum state invariant in ordinary measurement theory (Zeno etc). Commutation halts as the first reduction to the opposite boundary occurs. Self dies and re-incarnates at the opposite boundary. Negentropy gain results in general and can be see as the information gained in the computation. One might hope that the new self (maybe something at higher level of dark matter hierarchy) is a little bit wiser - at least statistically speaking this seems to be true by weak form of NMP!

4. One should understand the quantum counterparts for the basic rules of manipulation. ×,/,+, and - are the most familiar example.

    1. The basic rules correspond physically to generalized Feynman/twistor diagrams representing sequences of algebraic manipulations in the Yangian of super-symplectic algebra. Sequences correspond now to collections of partonic 2-surfaces defining vertices of generalized twistor diagrams.

    2. 3- vertices correspond to product and co-product for quantal stringy Noether charges. Geometrically the vertex - analog of algebraic operation - is a partonic 2-surface at with incoming and outgoing light-like 3-surfaces meet - like vertex of Feynman diagram. Co-product vertex is not encountered in simple algebraic systems, and is time reversed variant of vertex. Fusion instead of annihilation.

    3. This diagrammatics has a huge symmetry just like ordinary computations have. All computation sequences (note that the corresponding space-time surfaces are different!) connecting same collections A and B of objects produce the same scattering amplitude. This generalises the duality symmetry of hadronic string models. This is really gigantic simplification and the results of twistor Grassmann approach suggest that something similar is obtained there. This implication was so gigantic that I gave up the idea for years.

5. One should understand the analogs for the mathematical axioms. What are the fundamental rules of manipulation?

    1. The classical computation/deduction would obey deterministic rules at vertices. The quantal formulation cannot be deterministic for the simple reason that one has quantum non-determinism (weak form of NMP allowing also good and evil). The quantum rules obey the format that God used when communicating with Adam and Eve: do anything else but do not the break the conservation laws. Classical rules would list all the allowed possibilities and this leads to difficulties as Goedel demonstrated. I think that chess players follow the "anti-axiomatics".

    2. I have the feeling that anti-axiomatics (not any well-established idea, it occurred to me as I wrote this)could provide a more natural approach to quantum computation and even allow a new manner to approach to the

problematics of axiomatisations. It is also interesting to notice a second TGD inspired notion - the infinite hierarchy of mostly infinite integers (generated from infinite primes obtained by a repeated second quantization of an arithmetic QFT) - could make possible a generalisation of Gödel numbering for statements/computations. This view has at least one virtue: it makes clear how extremely primitive conscious entities we are in a bigger picture!

6. The laws of Physics take care that the anti-axioms are obeyed. Quite concretely:

    1. Preferred extremal property of Kähler action and Käler-Dirac action plus conservation laws for charges associated with super-symplectic and other generalised conformal symmetries would define the rules not broken in vertices.

    2. At the fermion lines connecting the vertices the propagator would be determined by the boundary part of Kahler-Dirac action. K-D equation for spinors and consistency consistency conditions from Kahler action (strong form of holography) would dictate what happens to fermionic oscillator operators defining the analog of quantum Boolean algebra as super-symplectic algebra.

posted by Matti Pitkanen @ <u>7:50 PM</u>

10 Comments:

At <u>4:22 AM</u>, Ⓑ <u>Ulla</u> said...

I must show my stupidity once again :P . One of the questions I have thought of lately is how a quantum number as aritmetic concept would look like, as instance inside a BH or as holography. If the 'computation' is made as something between boundaries this would be part of one boundary, but this is labelled by fractional aspects, that is a quantum phase. This decribes the uncertainty too, like a 'spagettified' number.

The complex numbers are a bit alike, but they are still ordered along a line. In a BH the 'numbers' are like fractional islands, but their displacement must still obey a law, as seen in the holography principle. This law must be a symmetry in my mind. It is like a quantum phase.

"Preferred extremal property of Kähler action and Käler-Dirac action plus conservation laws for charges associated with super-symplectic and other generalised conformal symmetries would define the rules not broken in vertices."

This statement talks of protected symmetries, like skyrmions, locally protected in hexagons. At least so, maybe even non-locally? In my mind it comes as a 'tunnelling' effect. I am still unsure about what that 'preferred' means, if it is a property of the symmetry itself?

If we look at symmetries they are also in a hierarchial order, with superradiant, as most non-local type, linear radiation, a dual and quasic structure. God herself? She is computating continously about every single detail that happen in the whole universe, in a holographic way, so her brain is a mirror of our own, also functions in the same way? This means our cortex and that symmetry are mirrors? Both has excellent supermemory :) through that symmetry.

If you can get a gist of what I say?

Look at this about quantum phase transitions. http://www.nature.com/ncomms/2015/150514/ncomms8111/full/ncomms8111.html

At 5:10 AM, Ulla said...

"By strong form of holography computations use only data at string world sheets and partonic 2-surfaces."

This can be compared to wavefunction and nodes in form of quasiparticles or Dirac fermions?

At 8:08 AM, Matpitka@luukku.com said...

I answer to the latter question first. Fermions in question are Dirac fermions. They are what I call fundamental fermions much like this in string model.

But fermions as a notion used in condensed matter physics are at a distance of light years conceptually. They emerge as approximate notion after the replacement of many-sheeted space-time with that of GRT and even that of special relativity- approximation as in Standard Model.

One forgets everything about string world sheets and partonic 2-sufaces as basic structures and many-sheeted space-time, and speaks about spinors in Minkowski space!

Quite a dramatic coarse graining but one crucial thing is common: symmetries and conservation laws, which are the fundamental truths respected in the physical axiomatics or should one say anti-axiomatics.

At <u>8:13 AM</u>, <u>Matpitka@luukku.com</u> said...

To Ulla:

I should not perhaps use the word 'God'. I try to use it with implicit ";-)". This because people associate with with all kinds of properties making him/her less abstract. By the way, the non-personal god could be NMP: principle rather than conscious entity;-).

Perhaps best manner to say this is to say that endless computation like process and and recreation is going on.

At <u>8:28 AM</u>, <u>Matpitka@luukku.com</u> said...

To Ulla:

Non-locality is present from beginning. The starting point of TGD was that particles corresponds to 3-dimensional surfaces rather than points.

The non-locallity is expressed by in Bell inequalities stating that particles in quantum mechanics have more correlations than they could have in any *purely local* theory relying on classical concept of probability (no interference of wave functions).

In TGD this fact has space-time correlates: strings connecting partonic 2-surfaces accompanied by flux tubes. TGD is a non-local theory but does not try to get rid of state function reduction like most the non-local theories such as Bohm's theory. TGD only removes un-necessary mystics- one might call it black magic- from quantum physics.

Einstein would be happy bout TGD (maybe he is ;-)-. Also Bohr would be relaxed snce he could reconsider whether it is really sensible to eliminate "ontology" from the vocabulary altogether and replace it with the attribute "crazy" as he did.

Here is no unique reality but this does not mean that there would be no reality. There are superpositions of quantum realities replaced by new ones all the time as this universe is studying itself by recreating itself again and again and storing its findings as negentropic entanglement - the Akashic records;-).

Ye, The Akashic records... the prime string...

Condensed matter does not tell things that coarsly, in my mind.

"They emerge as approximate notion after the replacement of many-sheeted space-time with that of GRT and even that of special relativity- approximation as in Standard Model." Exactly, and this is the beautiful part of it, y get the curvature.

The 3-surfaces comes from quarks? Like this? http://www.sciencedirect.com/science/article/pii/037594749190738R

GRT is approximation to relativity: relates like classical thermodynamics to quantum thermodynamics.

Quarks correspond to 3-surfaces as all particles do. The 3-surfaces define classical geometric and topological correlates for their quantal properties. In string models strings should provide this description but to be honest, they do not!

The link talks about something different. 3-quark systems, not 3-surfaces.

This comment has been removed by the author.

At <u>10:39 AM</u>,   <u>Ulla</u> said...

I must formulate again.

Ye, but mesons as diquarks then? Higgs particle is a dipole. Why would the foundation be a 3-surface and then go back to a dipole?

Color confinement is what you think of?

http://ptp.oxfordjournals.org/content/65/5/1684.full.pdf

The hadronic quark bag system?

How does a 3 surface define a topological correlate (should be a dipole in this 2D world)? Is it the stability, the simple fact that they are seen? Odd numbers, asymmetry? Partons as 2-surfaces are related how?

http://www.physics.gla.ac.uk/ppt/lqcd.htm

How is this a scalar field? Geometry and topology are made of vectors?

A link would be fine, I have searched your many papers for this. It is now important for me to know the detailed structure.

At <u>7:05 PM</u>, <u>Matpitka@luukku.com</u> said...

To Ulla: I answer the questions that I understand.

Topology is science of shape. Coffee cup and "munkkirinkila" are topologically the same thing. Geometry takes also distances into account, not only shape. Coffee cup and coffee cup upside down are geometrically one and same thing.

Scalar field could be seen as a notion of differential geometry. Differential geometry is purely local geometry. Angles between vectors for instance. But not shapes.

About 3-surface as topological correlate of particle. Consider first 2-D surfaces. Take a plane and add small handles. You can move these handles around. They are like particles, particles as topological inhomogenities in 2-D world. Generalize this to 3-D case and you get something that I mean. Reduction of particles to space-time topology.

Geometry made of vector is putting apple and orange in the same basket.

05/07/2015 - http://matpitka.blogspot.com/2015/05/breakthroughs-in-number-theoretic.html#comments

Breakthroughs in the number theoretic vision about TGD

Number theoretic universality states that besides reals and complex numbers also p-adic number fields are involved (they would provide the physical correlates of cognition). Furthermore, scattering amplitudes should be well-defined in all number fields be obtained by a kind of algebraic continuation. I have introduced the notion of intersection of realities and p-adicities which corresponds to some algebraic extension of rationals inducing an extension of p-adic numbers for any prime p. Adelic physics is a strong candidate for the realization of fusion of real and p-adic physics and would mean the replacement of real numbers with adeles. Field equations would hold true for all numer fields and the space-time surfaces would relate very closely to each other: one could say that p-adic space-time surfaces are cognitive representations of the real ones.

I have had also a stronger vision which is now dead. This sad event however led to a discovery of several important results.

1. The idea has been that p-adic space-time sheets would be not only "thought bubbles" representing real ones but also correlates for intentions and the transformation of intention to action would would correspond to a quantum jump in which p-adic space-time sheet is transformed to a real one. Alternatively, there would be a kind of leakage between p-adic and real sectors. Cognitive act would be the reversal of this process. It did not require much critical thought to realize that taking this idea seriously leads to horrible mathematical challenges. The leakage takes sense only in the intersection, which is number theoretically universal so that there is no point in talking about leakage. The safest assumption is that the scattering amplitudes are defined separately for each sector of the adelic space-time. This means enormous relief, since there exists mathematics for defining adelic space-time.

2. This realization allows to clarify thoughts about what the intersection must be. Intersection corresponds by strong form of holography to string world sheets and partonic 2-surfaces at which spinor modes are localized for several reasons: the most important reasons are that em charge must be well-defined for the modes and octonionic and real spinor structures can be equivalent at them to make possible twistorialization both at the level of imbedding space and its tangent space.

   The parameters characterizing the objects of WCW are discretized - that is belong to an appropriate algebraic extension of rationals so that surfaces are continuous and make sense in real number field and p-adic number fields. By conformal invariance they might be just conformal moduli. Teichmueller parameters, positions of punctures for partonic 2-surfaces, and corners and angles at them for string world sheets. These can be continued to real and p-adic sectors and

3. Fermions are correlates for Boolean cognition and anti-commutation relations for them are number theoretically universal, even their quantum variants when algebraic extension allows quantum phase. Fermions and Boolean cognition would reside in the number theoretically universal intersection. Of course they must do so since Boolean thought and cognition in general is behind all mathematics!

4. I have proposed this in p-adic mass calculations for two decades ago. This would be wonderful simplification of the theory: by conformal invariance WCW would reduce to finite-dimensional moduli space as far as calculations of scattering amplitudes are considered. The testing of the theory requires classical theory and 4-D space-time. This holography would not mean that one gives up space-time: it is necessary. Only cognitive and as it seems also fundamental sensory representations are 2-dimensional. All that one can mathematically say about reality is by using data at these 2-surfaces. The rest is needed but it require mathematical thinking and transcendence! This view is totally different from the sloppy and primitive philosophical idea that space-time could somehow emerge from discrete space-time.

This has led also to modify the ideas about the relation of real and p-adic physics.

1. The notion of p-adic manifolds was hoped to provide a possible realization of the correspondence between real and p-adic numbers at space-time level. It relies on the notion canonical identification mapping p-adic numbers to real in continuous

manner and realizes finite measurement resolution at space-time level. p-Adic length scale hypothesis emerges from the application of p-adic thermodynamics to the calculation of particle masses but generalizes to all scales.

2. The problem with p-adic manifolds is that the canonical identification map is not general coordinate invariant notion. The hope was that one could overcome the problem by finding preferred coordinates for imbedding space. Linear Minkowski coordinates or Robertson-Walker coordinates could be the choice for M^4. For CP_2 coordinates transforming linearly under U(2) suggest themselves. The non-uniqueness however persists but one could argue that there is no problem if the breaking of symmetries is below measurement resolution. The discretization is however also non-unique and makes the approach to look ugly to me although the idea about p-adic manifold as cognitive chargt looks still nice.

3. The solution of problems came with the discovery of an entirely different approach. First of all, realized discretization at the level of WCW, which is more abstract: the parameters characterizing the objects of WCW are discretized - that is assumed to belong to an appropriate algebraic extension of rationals so that surfaces are continuous and make sense in real number field and p-adic number fields.

Secondly, one can use strong form of holography stating that string world sheets and partonic 2-surfaces define the "genes of space-time". The only thing needed is to algebraically extend by algebraic continuation these 2-surfaces to 4-surfaces defining preferred extremals of Kähler action - real or p-adic. Space-time surface have vanishing Noether charges for a sub-algebra of super-symplectic algebra with conformal weights coming as n-ples of those for the full algebra- hierarchy of quantum criticalities and Planck constants and dark matters!

One does not try to map real space-time surfaces to p-adic ones to get cognitive charts but 2-surfaces defining the space-time genes to both real and p-adic sectors to get adelic space-time! The problem with general coordinate invariance at space-time level disappears totally since one can assume that these 2-surfaces have rational parameters. One has discretization in WCW, rather than at space-time level. As a matter fact this discretization selects punctures of partonic surfaces (corners of string world sheets) to be algebraic points in some coordinatization but in general coordinate invariant manner

4. The vision about evolutionary hierarchy as a hierarchy of algebraic extensions of rationals inducing those of p-adic number fields become clear. The algebraic extension associated with the 2-surfaces in the intersection is in question. The algebraic extension associated with them become more and more complex in evolution. Of course, NMP, negentropic entanglement (NE) and hierarchy of Planck constants are involved in an essential manner too. Also the measurement resolution characterized by the number of space-time sheets connecting average partonic 2-surface to others is a measure for "social" evolution since it defines measurement resolution.

There are two questions, which I have tried to answer during these two decades.

1. What makes some p-adic primes preferred so that one can say that they characterizes elementary particles and presumably any system?
2. What is behind p-adic length scale hypothesis emerging from p-adic mass calculations and stating that primes near but sligthly below two are favored physically, Mersenne primes in particular. There is support for a generalization of this hypothesis: also primes near powers of 3 or powers of 3 might be favored as length sand time scales which suggests that powers of prime quite generally are favored.

The adelic view led to answers to these questions. The answer to the first question has been staring directly to my eyes for more than decade.

1. The algebraic extension of rationals allow so called ramified primes. Rational primes decompose to product of primes of extension but it can happen that some primes of extension appear as higher than first power. In this case one talks about ramification. The product of ramified primes for rationals defines an integer characterizing the ramification. Also for extension allows similar characteristic. Ramified primes are an extremely natural candidate for preferred primes of an extension (I know that I should talk about prime ideals, sorry for a sloppy language): that preferred primes could follow from number theory itself I had not though earlier and tried to deduce them from physics. One can assign the characterizing integers to the string world sheets to characterize their evolutionary level. Note that the earlier heuristic idea that space-time surface represents a decomposition of integer is indeed realized in terms of holography!
2. Also infinite primes seem to find finally the place in the big picture. Infinite primes are constructed as an infinite hierarchy of second quantization of an arithmetic quantum field theory. The infinite primes of the previous level label the single fermion - and boson states of the new level but also bound states appear. Bound states can be mapped to irreducible polynomials of n-variables at n^{th} level of infinite obeying some restrictions. It seems that they are polynomials of a new variable with coefficients which are infinite integers at the previous level.

At the first level, bound state infinite primes correspond to irreducible polynomials: these define irreducible extensions of rationals and as a special case one obtains those satisfying so called Eistenstein criterion: in this case the ramified primes can be read directly from the form of the polynomial. Therefore the hierarchy of infinite primes seems to define algebraic extension of rationals, that of polynomials of one variables, etc.. What this means from the point of physics is a fascinating question. Maybe physicist must eventually start to iterate second quantization to describe systems in many-sheeted space-time! The marvellous thing would be the reduction of the construction of bound states - the really problematic part of quantum field theories - to number theory!

The answer to the second question requires what I call weak form of NMP.

1. Strong form of NMP states that negentropy gain in quantum jump is maximal: density matrix decompose into sum of terms proportional to projection operators: choose the sub-space for which number theoretic negentropy is maximal. The projection operator containing the largest power of prime is selected. The problem is that this does not allow free will in the sense as we tend to use: to make wrong choices!

2. Weak NMP allows to chose any projection operator and sub-space which is any sub-space of the sub-space defined by the projection operator. Even 1-dimensional in which case standard state function reduction occurs and the system is isolated from the environment as a prize for sin! Weak form of NMP is not at all so weak as one might think. Suppose that the maximal projector operator has dimension n_{max} which is product of large number of different but rather small primes. The negentropy gain is small. If it is possible to choose $n=n_{max}-k$, which is power of prime, negentropy gain is much larger!

It is largest for powers of prime defining n-ary p-adic length scales. Even more, large primes correspond to more refined p-adic topology: p=1 (one could call it prime) defines discrete topology, p=2 defines the roughest p-adic topology, the limit $p \rightarrow \infty$ is identified by many mathematicians in terms of reals. Hence large primes $p<n_{max}$ are favored. In particular primes near but below powers of prime are favored: this is nothing but a generalization of p-adic length scale hypothesis from p=2 to any prime p.

See the article What Could Be the Origin of Preferred p-Adic Primes and p-Adic Length Scale Hypothesis? For a summary of earlier postings see Links to the latest progress in TGD.

posted by Matti Pitkanen @ 9:08 PM

40 Comments:

At 3:34 AM, Anonymous said...

Hmm. So what about the phenomenal obviousness of ALL mathematical thinking and believing (in various number theories, axioms and axiomatics etc.) happening as "cognitive acts"?

At 6:37 AM, Matpitka@luukku.com said...

Looks sensible. Universe evolves more and more algebraically complex and becomes conscious of mathematical truths. There must be however something which cannot be reduced by the strong holography to 2-D basic objects (string world sheets and partonic 2-surfaces in the intersection of realities and p-adicities). We are are aware of space-time, were are aware of the idea of continuum, we are aware of completions of rationals and their extensions. This is something that goes beyond formulas.

At 8:03 PM, Stephen said...

What about these?
http://www.encyclopediaofmath.org/index.php?title=Wilson_polynomials

They have an interpretation as Racah coefficients for tensor products of irreducible representations of the group SU(2).

At 5:37 AM, Anonymous said...

http://plato.stanford.edu/entries/fictionalism-mathematics/

At 6:47 AM, Anonymous said...

Number theoretical universality of Boolean logic is highly questionable not only from fictionalist view but even inside the Platonist credo. In 'Sophist' Plato shows the codependent dialectics of fundamental categories ("sameness", "difference", "movement", "stillness", etc.), not unlike what Buddha and Buddhist logic says about codependent origins. So also in the Platonist approach, according to Plato himself, also geometry and number theory, not to mention classical bivalent Aristotelean/boolean logic, are not "outside" or "independent" of codependent origins.

All claims of all kinds of "universalities" can also be questioned from base of cultural anthropology and the study of human world views. In this approach, math based on bivalent boolean _functions_ is not cultural universal, but just one narrative that is not universally accepted even among mathematicians and logicians of Western culture.

Especially if they accept Gödel's proof that number theory of PM and those similar to it are incomplete, and cannot be consistently structured according to boolean or any other bivalent truth OR false logic.

Bohr's complementary philosophy is certainly not boolean, and allows both "true" and "false" as complementary "contradictions" and even as codependent views.

According to Bohr's interpretation, this non-universal Gödel-incomplete number theory (that schools and universities teach and indoctrinate in) is just part of the "classical world" that as a whole "measures" superposition. And as such, also number theory is ultimately a matter of choice...

At 6:51 AM, Anonymous said...

"As the Buddha may or may not have said (or both, or neither): 'There are only two mistakes one can make along the road to truth: not going all the way, and not starting.'"

The whole article is worth a read:

http://aeon.co/magazine/philosophy/logic-of-buddhist-philosophy/

At 5:36 AM, Matpitka@luukku.com said...

To anonymous about number theoretic universality etc... I cannot bat disagree. Godel's proof tells that complete axil systems are not possible: the pool of truth is infinitely deep. Something very easy to accept. Godel does say that Boolean logic is somehow wrong.   I see this as mathematicians, not anthropologist.

For me, number theoretical universality means just that it is superstructure behind all mathematics. This is simple fact. What is interesting is that it has concrete and highly non-trivial physical meaning when fermions are identified as physical correlates for Boolean thinking. Anticommutation relations for fermions are number theoretically universal and make sense in any number field. Bosonic commutation relations are not since they involve imaginary unit.

One can of course extend Boolean logic to quantum Boolean version and fermionic representations does this.

At 8:44 PM, Matpitka@luukku.com said...

Sorry for typo: Godel does **NOT** say that Boolean logic is somehow wrong.

Sorry for other typos too: these texting programs have their own vision about what I want to say. In any case: the essence of what I am saying is that fermionic anticommutation relations make sense in any number field - real or p-adic.

This is one facet of number theoretical universality and has a nice interpretation : one can construct endlessly variants of Boolean logic and call them logics but these constructions rely always on Boolean logic. You simply cannot think mathematically without Boolean logic. Fermions serving as correlates of Boolean logic define also a square root of geometry. This is a deep connection between logic and geometry: they really are the two basic pillars of existence.

Buddhist philosophy is nice but my goal is not to transform buddhist writings to mathematical statements. I regard myself as a scientist and for scientist authorities of any kind are poison. Kill the Buddha if you meet him on the road;-).

At 9:03 PM, Stephen said...

https://statistics.stanford.edu/sites/default/files/2000-05.pdf

see theorem 4.1, there's the 4 outcomes thing

0, -0.5, +0.5, 1

At 11:34 PM, Ulla said...

Gödels problem is an outcome of the 3 body problem if I have understood it right. This is also an effect of the 'collapse'?

2D quantum systems are always wider, bigger with more uncertainty, like some 'fuzzy logic'. Maybe the density differencies this brings along can be something to work on? At least now this is how I want to see the Hierarhchy of Plancks constant. Note how that density shifts when the Ouroborus bites its own tail :P Also the uncertainty shifts.

This is the essence of topology?

At <u>2:00 AM</u>, <u>Matpitka@luukku.com</u> said...

To Ulla:

Godel's theorem belongs meta-mathematics. It is difficult to imagine any physics application for it.

Effective 2-dimensionality means holography It might be seen as information theoretic notion 4-D space-time is redundant description if only scattering amplitudes are considered: 2-D basic objects are enough for this purpose. Classical correlates for quantum physics require 4-D space-time and one cannot throw it away.

At <u>8:11 AM</u>, Anonymous said...

In Boolean terms the proposition "Boolean thought and cognition in general is behind all mathematics!" is not just false, it is false!

At <u>7:20 PM</u>, <u>Matpitka@luukku.com</u> said...

Before making this kind of statements, we must define what "behind" means in the above statement. "Behind" of course does not mean that all mathematics reduces to Boolean algebra.This would be absolutely idiotic.

The natural meaning of "behind" is that <u>all</u> mathematics relies on deductions, which obey the rules of Boolean logic. p-Adic logic is not different from real one. The logic of differential geometry is not different from the logic of probability calculus.

If you can give example about a mathematical theory where deductions are carried out by using some other more romantic logic, please do so. I would be happy to get a concrete illustration about how my statement is false: I would prefer using Boolean logic. Just a concrete mathematical statement instead of something about Buddhism, egos, or anthropology. After than I am ready to consider seriously what you are claiming.

On the contrary, I believe the beauty of pure mathematics is very much in strict adherence to rigorous deductive logic based on clear definitions.

The freedom of choice in rigorously beautiful math is not in ad hoc axioms (to postulate set theory, real numbers etc. inherently illogical "completions of infinite processes"). The freedom of choice is at the foundational level, how we create and organize and deductively prove our number theory from a blank "zero state".

And you see that there are many areas of modern math,such as calculus etc. that throw away rigorous deductive logic and justify themselves by black magic alone, which also "works".

If you really prefer using rigorous deductive logic, you take side with Berkeley and Beauty and renounce the heretics Newton and Leibniz and their followers in the shadowy art of self-deception, without trying to escape into any kind of authoritarian argumentation (cf. "authoritarians rule academia", "black magic works", etc.).

Only after we agree to do also this bivalent deductive logic properly instead of by deception and black magic axiomatics, we can proceed to thinking about foundational level and "middle-way-math that would avoid all the following extremes:

a, b, both a and b, neither a or b.

As at least in the areas of math that involves measuring, boolean valuing does not work as things-events appear to be inherently approximate and fuzzy, and number theory based on middle-path-logic already at foundational level might be worth a thought.

At 4:23 AM, Matpitka@luukku.com said...

If you believe on rigorous deductive logic, then the first thing to do is to start to apply it. You can begin by justifying your claims by arguments. Saying "plainly wrong" without any justification is only an emotional rhetoric burst.

When you say inherently illogical "completions of infinite processes" you must tell what "inherently illogical" is. I find it very difficult to see what this phrase means when you talk about completions.

Mathematics without axioms is like Munchausen lifting himself into air. You simply cannot prove anything without axioms. They are a book keeping device for the what we believe to be truths. From a blank zero state you cannot deduce anything.

I like argumentation with contents. When some-one starts to talk about "self-deception", "authoritarian argumentation, "black magic"... I smell the presence of bad rhetoric.

Our observations about things are fuzzy, not things. I would be happy to see a number theory based on middle valued logic but I am afraid that this is only a nice sounding verbal construct like mathematics without axioms.

Quantum logic based on qubits is beautiful but logical statements makes only after the state function reduction yielding well-defined bits has happened. Our conscious experience makes the world classical.

At 5:27 AM, Anonymous said...

Then please, do your best to discuss the content. Even if the only content is your   confusion.   :)

"Infinite process" by definition means that the process (e.g. algorithm) continues ad infinitum, does not get finitely completed. I hope this clarifies what is illogical about "completions of infinite processes" in terms of binary logic. Based on this we can conclude that Boolean bivalent either-or valuing applies strictly only to finite phenomena, not to of infinite processes and approximations e.g. by some processes of limitations. Hence, any area of mathematics that deals with infinite processes in any way is not 'boolean' in the strict sense. Boolean thought and cognition can apply only to finite, bivalent processes that involve a Boolean identity.

"Axioms" are today used with variety of meanings. What was criticized was "ad hoc" axioms used to postulate what a mathematical physicist _wants_ to please himself with when rigorous deductive logic otherwise fails to produce the object of desire, not the foundational level axioms e.g. in Euclidean sense. Mathematics is a field of applied magic, and such use of ad hoc axioms is black magic.

"Our observations about things are fuzzy, not things." This is purely a statement of personal metaphysical belief, faith in "things" having "inherent ontology". That statement has nothing to do with logic, math and science.

Notions of 'length', 'area', 'volume', 'position', 'momentum' etc. are strictly speaking neither 'observations' nor 'things', but in this context just abstract mathematical notions, which in the mathematics we are used to do not behave in boolean way.

It is a pity that your comments are getting increasingly emotional and rhetoric. Mathematics probably looks black magic for anyone, who does not understand it. There is a lot of mathematics which looks black magic to me, but by working hardly I can get rid of this expression. I do not want to blame mathematics for my own limitations.

I will comment those parts of your comment which have some content.

*As mathematical structures, Boolean algebras extend without difficulty to continuous case: consider set theoretic realisation. p-Adics are completely well-defined notion and 2-adic number can be seen as infinite sequence if binary digits.

What is important that pinay digits are ordered: the higher the digit, the lower its significance. This is the basic idea of metric topology and makes it possible to work with continuum. Mathematics without continuum reduces to mere combinatorics.

*Completions of rationals to various number fields are a standard mathematical concept and extremely successful: if someone wants to believe that this notion is mathematically flawed, he can do so but certainly remains a lonely believer.

In any case, mathematicians discovered for centuries about that the notions of nearness, limit, continuity, Cauchy sequence and smoothness are extremely useful notions and allow to conclude the outcome of infinite processes. Very useful notions also for a philosopher of mathematics and highly recommended;-)

*Conscious logical thought- at least that part of which is representable physically - is discrete. Discreteness is one of the basic aspects of cognition - I formulate this in terms of cognitive resolution implying in turn finite measurement resolution.

*We should be be careful to not project the limitations of our cognition to the physical and mathematical realities. Materialists do this: they try to identify consciousness and physical reality and end up to a dead end.

At 7:48 AM, Anonymous said...

The only real argument presented above is "useful", which is purely emotional and rhetoric argument. "Useful" is exactly what is meant by "black magic", in contrast to the magic and beauty of rigorous mathematical deduction. Argument from authority is to refer what others think and do, instead of establishing a well defined theory of real numbers based on foundational axioms and Boolean chain of deductions and proves. If you want to play the game of Boolean mathematices, play it honestly and don't cheat at every corner.

In the Boolean context, it seemingly takes a complex self-deception to lose sight of the simple fact that there is indeed significant arithmetic symmetry break. As the basic carry rules of basic arithmetics state, Cauchy numbers in form ...nnn,p + ...nnn,p make arithmetic sense in the boolean context (ie, they can be added), but irrational Cauchy numbers n,nnn... + n,nnn... do not add up but remain vague and non-Boolean.

Unless, of course, you can prove that irrational Cauchy sequenses do add up in finite life time calculation in discrete non-vague manner and can be given a Boolean value. Go ahead, give it try.

I think it is time to stop the discussion since you are seem to be in rebel against mathematics and theoretical physics : windmills would be less dangerous enemy. People who argue that entire branches of science are totally wrong are usually called crackpots: I do not like that word because it is so much misused.

I have met many people who have declared war against some branch of eel-established science: one was logician who had the fix ide that special relativity contains logical errors: he had ended up with this conclusion by interpreting special relativity in Newtonian framework. I tried to explain but in vain.

You seems to misunderstand the idea of Cauchy sequence in a manner which remains for me black magic.

*Cauchy sequences provide a formulation of continuity: I fail to see how you manage to assign to them Boolean interpretation.

*You talk also about Cauchy numbers: by looking at Wikipedia you see that it is a dimensionless parameter used in hydrodynamics. I honesty admit that I fail to see the connection to the notion of continuity.

*Also the addition of Cauchy sequences is to my best knowledge completely irrelevant for the notion of limit.

At 9:28 AM, Matpitka@luukku.com said...

Some comments about axiomatics. This is of a technical tool for mathematicians. The unavoidable bureaucracy, one might say.

Theoretical physicist who is really constructing a theory is working hardly to find minimal number of basic assumptions, which might be true. Trying to find a set of assumptions which forms a coherent whole, is internally consistent, and predicts as much as possible.

This a process of trial and error and there is no point if declaring wars against mathematics or any other branch of science. This activity could not be farther from mechanical deduction from a fixed set of axioms, which requires just algorithms and in principle can be carried out by computer.

Theoretical physics has indeed led to powerful insight about mathematics: consider only Witten's work. Recently Nima Arkani Hamed and colleagues (mathematicians) have done similar job. This is by working as visionary: mathematicians take care of details when the dust has settled: this can take centuries.

Theoretician of course hopes that some day this structure can be axiomatized and even average professor can use the rules to calculate.

At 9:53 AM, Anonymous said...

No, it's not about politics, the question is very simple. Are "real numbers" numbers in boolean sense or combinatorical noise?

Present real numbers in the form of hindu-arabic cauchy sequenses in base 2. Pick a pair of such real numbers and add them up. Do you get a discrete result that starts with either one or zero?

AFAIK, no, and if not otherwise proven, hence "real numbers" cannot be said to be numbers in boolean sense.

And as real numbers cannot be said to be definable in boolean sense, that goes also for real complex plain, complex manifolds etc.

I don't know what the hell those things are, but they are certainly not "Boolean thought and cognition", presumably meaning numbers that can be expressed as either 1 or 0.

You can twist and dance around and play your politics and war games as much as you want - and it is sorry to see you try so hard not to admit what is so obvious -, but it's just math. This is just math, and if we choose to play Boolean game, we play it by Boolean rules, otherwise we would cheat.

Also in math you need to learn to walk before you can run. When you try to run before you have learned to take baby steps, you make a glorious short dash and then end up with your face in mud. And IMHO that summarizes the state of contemporary academic mathematics.

At <u>10:47 AM</u>, Anonymous said...

PS: IFF mathematics and theoretical physics claim to rule and conquer and control either openly or by implication, of course I revel, as any honest self-loving man would. :)

Thusly experiencing does not reduce to nor is limited by mathematics and theoretical physics, not even TGD. 2D representation of music is not same as picking a guitar in your lap and playing music that never was and never will be.

Anonymous, you are talking out of your ass. Stop wasting Matti's time

Now that we have hopefully left the dogmatic trenches or warfare and politics and are taking our baby steps, the structure of "real line" (-field) is as such an interesting object of study. In binary the sum of two points on real line (almost all of which are non-algebraic) is at least in most cases not:

0

1

both 0 and 1

neither 0 nor 1

but avoids all these extreme positions. ;)

So the notion of qubit seems now inherently related with adding elements of non-Boolean "completion" of rational line.

Also, it would seem that the bigger the number base, the smaller the room for vague. What would be the situation with base of Biggest Mersenne Known? Is there some kind or structural relation with "modern interpretation" of consistent histories and the questions it claims to allow and exclude?

see   <u>http://www.encyclopediaofmath.org/index.php/Algebra_of_sets</u>   for instance, its also called a σ-field σ-algebra, they can be unconditional, or conditional, upon all sorts of other spaces

go back and read the link I posted

Theorem 4.1

and quit babbling your wordy nonsense, Anonymous

random matrix theory alone cannot do it, the primes must enter in some way, and this spectral signature is universal for any and all unitary processes, apparently, if "U" know how to look , amirite M8? :)

Stephen, I have checked the link and now rechecked and found this gem under Proposition 5.1:

"Remark. From the equality, the infinite sum of squares converges to an _almost surely_ finite limit."

I humbly suggest that linguistic expression "almost surely" is pretty sure tell that the math in question has moved from the confines of Boolean values or "Boolean cognition" to somewhere else. ;)

Or maybe you can show that randomly picked real numbers in base 2, let's say 0,000... and 0,000... , without assuming that they are rationals, do really sum up in Boolean terms, ie. the sum is either a number beginning with 0 or 1, but not both or neither or something even more weird like "qubit"?

If you can't, we can't honestly say that "Boolean cognition" is behind claimed mathematical structures such as "real number line", and that the proof theory which is used to postulate such structures is a Boolean proof theory. And same goes for set theory.

https://statistics.stanford.edu/sites/default/files/2001-01.pdf is also very interesting "Unitary correlations and the Feijer kernel" is very interesting.

Your statement of "randomly picked real numbers in base 2" is ill-posed , are you just trying to reinvent some some "floating-point" representation ?

I'm saying its something more "weird" like a qubit.

nowhere in the thing u described do I see any sort of room for time, much less deterministic or nondeterministic notions of system states.

Fuzzy you say? nonsense, classical mechanics chaos, u take a Poincaire section of the flow and each orbit punctures that section at a particular point, repeat this many times (by dynamical system evaluation of integrals etc with whatever conditions) and you end up with a process whose "output" can take on a discrete number of values relative to the reference measure, well, basically, one can easily prove things such as Cantor, Levy 'dust' etc and go into fractal dimension which takes on any real value, so your argument is really just wildly stabbing in some direction or another, trying to talk big it seems...

this Boolean aspect of any set whatsoever requires the concept of indicator function, I(A)=1 if A in omega, 0 if its not in omega. integral. do you speak it? apparently not.

you sound like an I/T guy... are you an I/T guy? ;-)

See https://app.box.com/files/0/f/0/1/f_30073473869 for some interesting ways that numbers 7 and 24 just pop out of some rather elementary integrals

I do not want to use my time to ponder whether there is some conspiracy of power greedy mathematicians and physicists against civilized world. I just want to make clear some elementary things about Cauchy sequences in hope that they remove the feeling of black magic.

a) Cauchy sequences are used by everyone who can sum, subtract, multiply and divide decimal numbers. These are special kind of Cauchy sequences in which n^{th} term is the number in approximation using n decamical digits. One can use also binary or binary and much more general sequences.

These particular sequences are however convenient since all arithmetic operations are for rational numbers.

b) In numerics on introduces decimal/pinary/... cutoff. This makes sense if the functions are continuous and the operations for them respect continuity.

c) If one wants to formulate this axiomatically one can say that one works in the category of continuous functions. Absolutely no crime against mankind is involved. Everything is finite and numbers of operations are finite but approximate with an error that can be estimated. Computers use routinely binary Cauchy sequences with a success.

c) One could of course throw away real numbers as a conspiracy against mankind and decides to use only rationals (I do not know whether algebraic numbers are also doomed to to be part of conspiracy) . This leads to difficulties.

One must define differential equations etc as difference equations by specifying the size of different: single equation would be replaced by infinite number of them- one for each accuracy. Calculus and most of what has been achieved since Newton would be lost since no-one wants to write Newton's mechanics or Maxwell's theory or quantum field theory using difference equations: it would incredibly clumsy.

There would no exponent function, no Gaussian, no pi, no special functions. Things become in practice impossible. Most of number theory is lost: forget Riemann Zeta, forget p-adic numbers, ... Analytic calculations absolutely central in all science would become impossible.

Reals represent the transcendent, spirituality, going beyond what we can represent by counting with fingers. Recent science is deeply spiritual and in very concrete manner but the materialistic dogma prevents us from seeing this.

At 7:35 PM, Matpitka@luukku.com said...

I hope that the importance of the notion of finite accuracy became clear. It certainly does not look like a beautiful notion in its recent formulation.

Finite accuracy is the counterpart for finite measurement resolution/cognitive resolution and is anotion, which is often not considered explicitly in math text books. It is fundamental in physics but the problem is how to formulate it elegantly.

It is also encountered in in the adelic vision based on strong form of holography. One can in principle deduce scattering amplitudes in an algebraic extension of rationals (this for the parameters such as momenta appearing in them). One can algebraically continue this expression to all number fields.

But what if one cannot calculate the amplitudes exactly in the algebraic extension? There is no problem in real topology using ordinary continuity. But continuation to p-adic topologies is difficult since even a smallest change in rational number in real sense can mean very big change in p-adic sense. It seems that one cannot avoid canonical identification or some of its variants if one wants to assign to a real amplitude a p-adic amplitude in continuous manner.

Finite accuracy is also a deep physical concept: fermions at string world sheets are Boolean cognitive representation of space-time geometry. But in finite accuracy representing 4-D object using data assignable to a collection of 2-D objects rather than 0-dimensional objects (points) as in the usual naive discretization, which is not consistent with symmetries. Discrete set of points is replaced with discrete collection of 2-surfaces labelled by parameters in algebraic extension of rationals. The larger the density of strings, the better the representation. This is strong form or holography is implied by strong form of general coordinate invariance: a completely unexpected connection to Einstein's great principle.

This leads also to an elegant realization of number theoretical universality and hierarchy of inclusions of hyper-finite factors as a realization of finite measurement resolution. Also evolution as increase of complexity of algebraic extension of rationals pops up labelled by integers n =h_eff/h, which are products of ramified primes characterizing the sub-algebra of super-symplectic algebra acting as gauge conformal gauge symmetries. Effective Planck constant has purely number theoretic meaning as a measure for the complexity of algebraic extension!

Ramification is also number theoretic correlate of quantum criticality! Rational prime decomposes to product of prime powers such that some of them are higher than first powers: analog for multiple root in polynomial - criticality! For me this looks amazingly beautiful.

At 7:43 PM, Matpitka@luukku.com said...

Correction to the last paragraph: "prime powers such that" should read "powers of primes of extension such that"

At 5:08 AM, Anonymous said...

I don't know what all will be lost if we honestly admit that "real numbers" do not behave arithmetically, at least in the boolean sense, and though many say that "real numbers satisfy the usual rules of arithmetic", obviously they don't. Any child can see that emperor has no clothes in that respect.

Even though reals don't, AFAIK the p-adic side does satisfy the usual rules of arithmetic, at least in some areas. Worth a more careful look. Cauchy intervals within intervals is perfectly OK and very rich and interesting structure, and repeating patterns of rationals is amazing and beautiful thing worth deeper study, e.g. how do lengths of repeating patterns behave in various bases, on both sides of rationals? When repeating patterns are plotted inside Cauchy intervals, I see a wave forms at very basic level of number theory.

In OP Matti does see the light saying that mathematical structures follow from number theory itself, trying to deduce from "physics" does not work.

So here is relatively simple question: what is the _minimum_ of number theory you need to observe quantum observables? I'm very much in doubt that e.g. "canonical identification" is needed (but rather, confuses and messes things up).

I'm not I/T, but even I know that computers don't do real numbers or any other infinite processes. Floating points, lazy algorithms, etc. get the job done. Finite accuracy works in boolean way, but no, we can't say that "finite accuracy" strings are "real numbers".

If we insist that problem of mathematical continuum "has been solved with "least upper bound" completion of algebraic (e.g. roots) and algorithmic (pi, e) realside infinite processes", there is a cost: the solution is not boolean, rules of arithmetic don't work regardless of how much some people pretend that they work and push the problems under the mattress and out of text books. It's not about politics, it's just math.

There are other options, we can admit that the problem of mathematical continuum remains unsolved, or poorly understood and formulated, and keep on thinking and questioning, instead of blindly believing the academic authorities that say that real numbers follow the basic rules of arithmetics. Eppur si muove.

At 5:29 AM, Matpitka@luukku.com said...

This is my last comment in this fruitless discussion. I have done pedagogical efforts in order to clarify the basics but in vain. I however strongly encourage to continue serious studies of basics before declaring a war against modern mathematics and physics.

I have tried to explain that finite calculational accuracy is the point: it is not possible to calculate real number exactly in finite time and no-one has been claiming anything like that. The idea of giving up all the mathematics since Newton is just just because we cannot calculate with infinite precision is complete idiotism.

And I am still unable to see what is wrong with Cauchy sequences: here I tried to concretise them in terms of decimal representation in order to give the idea what they are about but it seems that it did not help.

The generalisation of real numbers rather than refusing to admit their existence, is the correct direction to proceed and I have been working with this problem with strong physical motivations. Fusion of reals and p-adics to adelic structures also at space-time level, hierarchy of infinite primes defining infinite hierarchy of second quantisation for an arithmetic quantum field theory, even construction of arithmetics of Hilbert spaces, replacement of real point with infinitely structured point realizing number theoretic Brahman = Atman/algebraic holography. These are interesting lines to proceed rather than a return to cave.

Strange that someone blames me for blindly believing academic authorities;-).

As for relation of Boolean operators V and its vertical inverse, and human cognition, propositional logic is far from universal; some natural languages behave closer to propositional logic, some not in the slightest.

Leveled horizontal operators of ordinality "<" and ">" (less-more) are much more naturally universal in human cognition, I'm not aware of natural language without more-less relation, which is also naturally hyperfinite process closely related to whole-part relation. The arrows giving directions are also more-less relations: go more in the direction the arrow is pointing, less in the opposite direction. These operators predate all other written language and needless to say, propositional logic.

Matti, read again. Your latest comment has very little to do with what has been said and meant.

Again: The authorities (e.g. wiki) keep on saying that real numbers follow the basic rules of arithmetics. Obviously that claim is not true.

The definition of 'real number' refers to infinite process ("least upper bound"), not to finite computable segment. Finite segments by definition are NOT "real numbers", they are something else. Some say "approximations", but also an approximation is NOT a real number. It is an approximation.

If we want to keep math communicable, we must respect definitions and do our best to define as clearly as we can. The notion of "real number" is as it is usually used, horribly vague and poorly defined.

That is of course a big IF and communication is not necessarily priority. The word "sin" has been mentioned in context of incommunicado.

At 2:17 PM, Stephen said...

Anonymous, who is this we you speak of? get some books and stop watching videos. do some analysis. read up on Baire spaces if u are so caught up on notions of continuity

At 2:24 PM, Stephen said...

for example, ur little toy problem: The irrational numbers, with the metric defined by , where is the first index for which the continued fraction expansions of a and b differ (this is a complete metric space).

the iterated map that gives rise to the continted fraction expansion of a real number .. well, it's related to the riemann zeta function, see the Wikipedia page

Edit

Continued fractions also play a role in the study of dynamical systems, where they tie together the Farey fractions which are seen in the Mandelbrot set with Minkowski's question mark function and the modular group Gamma.

The backwards shift operator for continued fractions is the map $h(x) = 1/x - \lfloor 1/x \rfloor$ called the Gauss map, which lops off digits of a continued fraction expansion: $h([0; a_1, a_2, a_3, \ldots]) = [0; a_2, a_3, \ldots]$. The transfer operator of this map is called the Gauss–Kuzmin–Wirsing operator. The distribution of the digits in continued fractions is given by the zero'th eigenvector of this operator, and is called the Gauss–Kuzmin distribution.

At <u>6:19 PM</u>, ☀ <u>Matpitka@luukku.com</u> said...

To Anonymous: You should calm down and stop talking total nonsense.

You are unable to realize that things can exist although we cannot know perfectly what they are. What we can know is that real number is in some segment, we can narrow down this segment endlessly but never know exactly.

But we have also something else than mere numerical computation: we have the conscious intelligence. It cannot be computerised or axiomatised but and most importantly, it is able to discover new truths.

In mathematics communication requires also learning: just watching some videos and becoming a fan of some character making strong nonsense claims is not enough. Also mathematical intuition is something very difficult to teach: some of us have it, others do not.

Just as some people are able to compose marvellous music. It seems that we must just accept this. I am not musically especially talented but I enjoy the music of the great composers and experience the miracle again and again: I do not declare war against this music.

At <u>7:17 PM</u>, ☀ <u>Matpitka@luukku.com</u> said...

Some comments about quantum Boolean thinking and computation as I see it to happen at fundamental level.

a) One should understand how Boolean statements A-->B are represented. Or more generally, how a computation like procedure leading from a collection A of math objects collection B of math objects takes place. Recall that in computations the objects coming in and out are bit sequences. Now one have computation like process. --> is expected to correspond to the arrow of time.

If fermionic oscllator operators generate Boolean basis, zero energy ontology is necessity to realize rules as rules connecting statements realized as bit sequences. Positive energy ontology would allow only statements. Collection A is at the lower passive boundary of CD and B at the upper active one. As a matter fact, it is a quantum superpositions of Bs, which is there! In the quantum jump selecting single B at the active boundary, A is replaced with a superposition of A:s: self dies and re-incarnates and generates negentropy. Q-computation halts.

That both a and b cannot be known precisely is a quantal limitation to what can be known: philosopher would talk about epistemology here. The different pairs (a,b) in superposition over b:s are analogous to different implications of a. Thinker is doomed to always live in a quantum cognitive dust and never be quite sure of.

Continuing....

b) What is the computer like structure now? Turing computer is 1-D time-like line. This quantum computer is superposition of 4-D space-time surfaces with the basic computational operations located along it as partonic 2-surfaces defining the algebraic operations and connected by fermion lines representing signals. Very similar to ordinary computer.

c) One should understand the quantum counterparts for the basic rules of manipulation. x,/,+, and - are the most familiar example.

*The basic rules correspond physically to generalized Feynman/twistor diagrams representing sequences of algebraic manipulations in the Yangian of super-symplectic algebra. Sequences correspond now to collections of partonic 2-surfaces defining vertices of generalized twistor diagrams.

*3- vertices correspond to product and co-product represented as stringy Noether charges. Geometrically the vertex - analog of algebraic operation - is a partonic 2-surface at with incoming and outgoing light-like 3-surfaces meet - like vertex of Feynman diagram. There is also co-product vertex not encountered in simple algebraic systems, it is time reversed variant of vertex. Fusion instead of annihilation.

*There is a huge symmetry as in ordinary computations too. All computation sequences connecting same collections A and B of objects produce the same scattering amplitude. This generalises the duality symmetry of hadronic string models. This is really gigantic simplification and the results in twistor program suggest that something similar is obtained there. This implication was so gigantic that I gave up the idea for years.

d) One should understand the analogs for the mathematical axioms. What are the fundamental rules of manipulation?

*The classical computation/deduction would obey deterministic rules at vertices. The quantal formulation cannot be deterministic for the simple reason that one has quantum non-determinism (weak form of NMP allowing also good and evil) . The quantum rules obey the format that God used when communicating with Adam and Eve: do anything else but do not the break conservation laws. Classical rules would list all the allowed possibilities and this leads to difficulties as Goedel demonstrated. I think that chess players follow the anti-axiomatics.

I have the feeling that anti-axiomatics could give a more powerful approach to computation and deduction and allow a new manner to approach to the problematics of axiomatisations. Note however that the infinite hierarchy of mostly infinite integers could make possible a generalisation of Godel numbering for statements/computations.

e) The laws of physics take care that the anti-axioms are obeyed. Quite concretely:

*Preferred extremal property of Kaehler action and Kaeler-Dirac action plus conservation laws for charges associated with super-symplectic and other generalised conformal symmetries would define the rules not broken in vertices.

*At the fermion lines connecting the vertices the propagator would be determined by the boundary part of Kahler-Dirac action. K-D equation for spinors and consistency consistency conditions from Kahler action (strong form of holography) would dictate what happens to fermionic oscillator operators defining the analog of quantum Boolean algebra as super-symplectic algebra.

05/05/2015 - http://matpitka.blogspot.com/2015/05/updated-negentropy-maximization.html#comments

Updated Negentropy Maximization Principle

Quantum-TGD involves "holy trinity" of time developments. There is the geometric time development dictated by the preferred extremal of Kähler action crucial for the realization of General Coordinate Invariance and analogous to Bohr orbit. There is what I originally called unitary "time development" $U: \Psi_i \to U\Psi_i \to \Psi_f$, associated with each quantum jump. This would be the counterpart of the Schrödinger time evolution $U(-t, t \to \infty)$. Quantum jump sequence itself defines what might be called subjective time development.

Concerning U, there is certainly no actual Schrödinger equation involved: situation is in practice same also in quantum field theories. It is now clear that in Zero Energy Ontology (ZEO) U can be actually identified as a sequence of basic steps such that single step involves a unitary evolution inducing delocalization in the moduli space of causal diamonds CDs) followed by a localization in this moduli space selecting from a superposition of CDs single CD. This sequence replaces a sequence of repeated state function reductions leaving state invariant in ordinary QM. Now it leaves in variant second boundary of CD (to be called passive boundary) and also the parts of zero energy states at this boundary. There is now a very attractive vision about the construction of transition amplitudes for a given CD, and it remains to be see whether it allows an extension so that also transitions involving change of the CD moduli characterizing the non-fixed boundary of CD.

A dynamical principle governing subjective time evolution should exist and explain state function reduction with the characteristic one-one correlation between macroscopic measurement variables and quantum degrees of freedom and state preparation process. Negentropy Maximization Principle is the candidate for this principle. In its recent form it brings in only a single little but overall important modification: state function reductions occurs also now to an eigen-space of projector but the projector can now have dimension which is larger than one. Self has free will to choose beides the maximal possible dimension for this sub-space also lower dimension so that one can speak of weak form of NMP so that negentropy gain can be also below the maximal possible: we do not live in the best possible world. Second important ingredient is the notion of negentropic entanglement relying on p-adic norm.

The evolution of ideas related to NMP has been slow and tortuous process characterized by misinterpretations, over-generalizations, and unnecessarily strong assumptions, and has been basically evolution of ideas related to the anatomy of quantum jump and of Quantum-TGD itself.

Quantum measurement theory is generalized to theory of Consciousness in TGD framework by replacing the notion of observer as outsider of the physical world with the notion of self. Hence it is not surprising that several new key notions are involved.

1. ZEO is in central role and brings in a completely new element: the arrow of time changes in the counterpart of standard quantum jump involving the change of the passive boundary of CD to active and vice versa. In living matter the changes of the of time are inn central role: for instance, motor action as volitional action involves it at some level of self hierarchy.
2. The fusion of real physics and various p-adic physics identified as physics of cognition to single adelic physics is second key element. The notion of intersection of real and p-adic worlds (intersection of sensory and cognitive worlds) is central and corresponds in recent view about TGD to string world sheets and partonic 2-surfaces whose parameters are in an algebraic extension of rationals. By strong form of of holography it is possible to continue the string world sheets and partonic 2-surfaces to various real and p-adic surfaces so that what can be said about quantum physics is coded by them. The physics in algebraic extension can be continued to real and various p-adic sectors by algebraic continuation meaning continuation of various parameters appearing in the amplitudes to reals and various p-adics.

An entire hierarchy of physics labeled by the extensions of rationals inducing also those of p-adic numbers is predicted and evolution corresponds to the increase of the complexity of these extensions. Fermions defining correlates of Boolean cognition can be said so reside at these 2-dimensional surfaces emerging from strong form of holography implied by strong form of general coordinate invariance (GCI).

An important outcome of adelic physics is the notion of number theoretic entanglement entropy: in the defining formula for Shannon entropy logarithm of probability is replaced with that of p-adic norm of probability and one assumes that the p-adic prime is that which produces minimum entropy. What is new that the minimum entropy is negative and one can speak of negentropic entanglement (NE). Consistency with standard measurement theory allows only NE for which density matrix is n-dimensional projector.

3. Strong form of NMP states that state function reduction corresponds to maximal negentropy gain. NE is stable under strong NMP and it even favors its generation. Strong form of NMP would mean that we live in the best possible world, which does not seem to be the case. The weak form of NMP allows self to choose whether it performs state function reduction yielding the maximum possible negentropy gain. If n-dimensional projector corresponds to the maximal negentropy gain, also reductions to sub-spaces with n-k-dimensional projectors down to 1-dimensional projector are possible. Weak form has powerful implications: for instance, one can understand how primes near powers of prime are selected in evolution identified at basic level as increase of the complexity of algebraic extension of rationals defining the intersection of realities and p-adicities.

4. NMP gives rise to evolution. NE defines information resources, which I have called Akashic records (a kind of Universal library). The simplest possibility is that under the repeated sequence of state function reductions at fixed boundary of CD NE at that boundary becomes conscious and gives rise to experiences with positive emotional coloring: experience of love, compassion, understanding, etc... One cannot exclude the possibility that NE generates a conscious experience only via the analog of interaction free measurement but this option looks un-necessary in the recent formulation.

5. Dark matter hierarchy labelled by the values of Planck constant $h_{eff}=n\times h$ is also in central role and interpreted as a hierarchy of criticalities in which sub-algebra of super-symplectic algebra having structure of conformal algebra allows sub-algebra acting as gauge conformal algebra and having conformal weights coming as n-ples of those for the entire algebra. The phase transition increasing h_{eff} reduces criticality and takes place spontaneously. This implies a spontaneous generation of macroscopic quantum phases interpreted in terms of dark matter. The hierarchies of conformal symmetry breakings with n(i) dividing n(i+1) define sequences of inclusions of HFFs and the conformal sub-algebra acting as gauge algebra could be interpreted in terms of measurement resolution.

n-dimensional NE is assigned with $h_{eff}=n\times h$ and is interpreted in terms of the n-fold degeneracy of the conformal gauge equivalence classes of space-time surfaces connecting two fixed 3-surfaces at the opposite boundaries of CD: this reflects the non-determinism accompanying quantum criticality. NE would be between two dark matter systems with same h_{eff} and could be assigned to the pairs formed by the n sheets. This identification is important but not well enough understood yet. The assumption that p-adic primes p divide n gives deep connections between the notion of preferred p-adic prime, negentropic entanglement, hierarchy of Planck constants, and hyper-finite factors of type II_1.

6. Quantum-Classical correspondence (QCC) is an important constraint in ordinary measurement theory. In TGD, QCC is coded by the strong form of holography assigning to the quantum states assigned to the string world sheets and partonic 2-surfaces represented in terms of super-symplectic Yangian algebra space-time surfaces as preferred extremals of Kähler action, which by quantum criticality have vanishing super-symplectic Noether charges in the sub-algebra characterized by integer n. Zero modes, which by definition do not contribute to the metric of "world of classical worlds" (WCW) code for non-fluctuacting classical degrees of freedom correlating with the quantal ones. One can speak about entanglement between quantum and classical degrees of freedom since the quantum numbers of fermions make themselves visible in the boundary conditions for string world sheets and their also in the structure of space-time surfaces.

NMP has a wide range of important implications.

1. In particular, one must give up the standard view about second law and replace it with NMP taking into account the hierarchy of CDs assigned with ZEO and dark matter hierarchy labelled by the values of Planck constants, as well as the effects due to NE. The breaking of second law in standard sense is expected to take place and be crucial for the understanding of evolution.
2. Self hierarchy having the hierarchy of CDs as imbedding space correlate leads naturally to a description of the contents of consciousness analogous to thermodynamics except that the entropy is replaced with negentropy.
3. In the case of living matter NMP allows to understand the origin of metabolism. NMP demands that self generates somehow negentropy: otherwise a state function reduction to tjhe opposite boundary of CD takes place and means death and re-incarnation of self. Metabolism as gathering of nutrients, which by definition carry NE is the manner to avoid this fate. This leads to a vision about the role of NE in the generation of sensory qualia and a connection with metabolism. Metabolites would carry NE and each metabolite would correspond to a particular qualia (not only energy but also other quantum numbers would correspond to metabolites). That primary qualia would be associated with nutrient flow is not actually surprising!
4. NE leads to a vision about cognition. Negentropically entangled state consisting of a superposition of pairs can be interpreted as a conscious abstraction or rule: negentropically entangled Schrödinger cat knows that it is better to keep the bottle closed.
5. NMP implies continual generation of NE. One might refer to this ever expanding universal library as "Akaschic records". NE could be experienced directly during the repeated state function reductions to the passive boundary of CD - that is during the life cycle of sub-self defining the mental image. Another, less feasible option is that interaction free measurement is required to assign to NE conscious experience. As mentioned, qualia characterizing the metabolite carrying the NE could characterize this conscious experience.
6. A connection with fuzzy qubits and quantum groups with NE is highly suggestive. The implications are highly non-trivial also for quantum computation allowed by

weak form of NMP since NE is by definition stable and lasts the lifetime of self in question.

For details see the chapter Negentropy Maximization Principle of "TGD Inspired Theory of Consciousness".   For a summary of the earlier postings see Links to the latest progress in TGD.

posted by Matti Pitkanen @ 5:09 AM

6 Comments:

At 9:53 AM, Anonymous said...

Matti, I know this is a "narrow specialization" but (stochastic) point processes are actually related to the Ising 'problem' and Riemann zeta hypothesis and fermions/bosons.

See
https://books.google.com/books?id=3Iw2NTPsS6QC&pg=PA9&lpg=PA9&dq=point+processes+1950-2000+vere-jones&source=bl&ots=DNgQ0dIbGO&sig=w285gr6bA6jeNhD7XlqcXeJ067s&hl=en&sa=X&ei=kXBKVYn6IIXfsATlh4HQBA&ved=0CCQQ6AEwAA#v=onepage&q=point%20processes%201950-2000%20vere-jones&f=false

see page 14, Diaconis and Evans, 2001 and Coram and Diaconis 2002

can you please comment?

--crow

At 10:13 AM, Anonymous said...

the referenced article is available in pdf format at http://www.sciencedirect.com/science/article/pii/S0097316500930978 the other article is at http://statweb.stanford.edu/~cgates/PERSI/papers/coram03.pdf

--crow

At 6:53 PM, Matpitka@luukku.com said...

To Crow:

I would be happy say something interesting about these articles. Unfortunately I cannot. This kind statistical approach I have not been working with.

At <u>10:05 PM</u>, <u>Stephen</u> said...

Brahman is full of all perfections. And to say that Brahman has some purpose in creating the world will mean that it wants to attain through the process of creation something which it has not. And that is impossible. Hence, there can be no purpose of Brahman in creating the world. The world is a mere spontaneous creation of Brahman. It is a Lila, or sport, of Brahman. It is created out of Bliss, by Bliss and for Bliss. Lila indicates a spontaneous sportive activity of Brahman as distinguished from a self-conscious volitional effort. The concept of Lila signifies freedom as distinguished from necessity.

—Ram Shanker Misra, The Integral Advaitism of Sri Aurobindo

http://en.m.wikipedia.org/wiki/Lila_(Hinduism)

Surely has some the TGD cognate and also ti axiom of choice in math somehow

At <u>1:29 PM</u>, <u>Stephen</u> said...

Matti, I just realized my links that I posted in the comments on your other post are probably more appropriate for this post, in regards to "unitary time development"

https://statistics.stanford.edu/sites/default/files/2001-01.pdf

see the title "unitary correlations and the Fejer kernel" irreducible group characters and whatnot

At <u>6:05 PM</u>, <u>Matpitka@luukku.com</u> said...

The recent view is that unitary evolution would correspond in TGD a dispersion in the space of moduli characterising the position of the upper boundary of CD and would also affect the states at the upper boundary.

Unfortunately, I am not able to say much about this. Sequence of steps: U followed by localisation in moduli.

Concerning zeros of zeta, I take again seriously the hypothesis that they might correspond to conformal weights of supersymplectic algebra characterising exponents of powers of radial light- like coordinate. Also this algebra as a fractal hierarchy of sub-algebras for which are n-ples of those for the full algebra. The number of generators is infinite: the zeros of zeta! In case of ordinary Kac-Moody type algebras generators have only finite number of different weights. This would be something incredibly complex.

Conformal confinement would be implication: physical states would have real conformal weights: the sum over multiples of imaginary parts of Riemann zeros would vanish for them.

04/29/2015 - http://matpitka.blogspot.com/2015/04/what-could-be-origin-of-p-adic-length.html#comments

What could be the origin of p-adic length scale hypothesis?

The argument would explain the existence of preferred p-adic primes. It does not yet explain p-adic length scale hypothesis stating that p-adic primes near powers of 2 are favored. A possible generalization of this hypothesis is that primes near powers of prime are favored. There indeed exists evidence for the realization of 3-adic time scale hierarchies in living matter (see this) and in music both 2-adicity and 3-adicity could be present, this is discussed in TGD inspired theory of music harmony and genetic code (see this).

The weak form of NMP might come in rescue here.

1. Entanglement negentropy for a negentropic entanglement characterized by n-dimensional projection operator is the $\log(N_p(n))$ for some p whose power divides n. The maximum negentropy is obtained if the power of p is the largest power of prime divisor of p, and this can be taken as definition of number theoretic entanglement negentropy. If the largest divisor is p^k, one has $N= k\times \log(p)$. The entanglement negentropy per entangled state is $N/n=k\log(p)/n$ and is maximal for $n=p^k$. Hence powers of prime are favoured which means that p-adic length scale hierarchies with scales coming as powers of p are negentropically favored and should be generated by NMP. Note that $n=p^k$ would define a hierarchy of $h_{eff}/h=p^k$. During the first years

of h_{eff} hypothesis I believe that the preferred values obey $h_{eff}=r^k$, r integer not far from r= 2^{11}. It seems that this belief was not totally wrong.

2. If one accepts this argument, the remaining challenge is to explain why primes near powers of two (or more generally p) are favoured. $n=2^k$ gives large entanglement negentropy for the final state. Why primes $p=n_2= 2^k-r$ would be favored? The reason could be following. $n=2^k$ corresponds to p=2, which corresponds to the lowest level in p-adic evolution since it is the simplest p-adic topology and farthest from the real topology and therefore gives the poorest cognitive representation of real preferred extremal as p-adic preferred extermal (Note that p=1 makes formally sense but for it the topology is discrete).

3. Weak form of NMP suggests a more convincing explanation. The density matrix of the state to be reduced is a direct sum over contributions proportional to projection operators. Suppose that the projection operator with largest dimension has dimension n. Strong form of NMP would say that final state is characterized by n-dimensional projection operator. Weak form of NMP allows free will so that all dimensions n-k, k=0,1,...n-1 for final state projection operator are possible. 1-dimensional case corresponds to vanishing entanglement negentropy and ordinary state function reduction isolating the measured system from external world.

4. The negentropy of the final state per state depends on the value of k. It is maximal if n-k is power of prime. For $n=2^k=M_k+1$, where M_k is Mersenne prime n-1 gives the maximum negentropy and also maximal p-adic prime available so that this reduction is favoured by NMP. Mersenne primes would be indeed special. Also the primes $n=2^k-r$ near 2^k produce large entanglement negentropy and would be favored by NMP.

5. This argument suggests a generalization of p-adic length scale hypothesis so that p=2 can be replaced by any prime.

This argument together with the hypothesis that preferred prime is ramified would correlate the character of the irreducible extension and character of super-conformal symmetry breaking. The integer n characterizing super-symplectic conformal sub-algebra acting as gauge algebra would depends on the irreducible algebraic extension of rational involved so that the hierarchy of quantum criticalities would have number theoretical characterization. Ramified primes could appear as divisors of n and n would be essentially a characteristic of ramification known as discriminant.

An interesting question is whether only the ramified primes allow the continuation of string world sheet and partonic 2-surface to a 4-D space-time surface. If this is the case, the assumptions behind p-adic mass calculations would have full first principle justification.

For details see the article The Origin of Preferred p-Adic Primes? For a summary of earlier postings see Links to the latest progress in TGD.

How preferred p-adic primes could be determined?

p-Adic mass calculations allow to conclude that elementary particles correspond to one or possible several preferred primes assigning p-adic effective topology to the real space-time sheets in discretization in some length scale range. TGD inspired theory of consciousness leads to the identification of p-adic physics as physics of cognition. The recent progress leads to the proposal that Quantum-TGD is adelic: all p-adic number fields are involved and each gives one particular view about physics.

Adelic approach plus the view about evolution as emergence of increasingly complex extensions of rationals leads to a possible answer to th question of the title. The algebraic extensions of rationals are characterized by preferred rational primes, namely those which are ramified when expressed in terms of the primes of the extensions. These primes would be natural candidates for preferred p-adic primes.

1. Earlier attempts

How the preferred primes emerges in this framework? I have made several attempts to answer this question.

1. Classical non-determinism at space-time level for real space-time sheets could in some length scale range involving rational discretization for space-time surface itself or for parameters characterizing it as a preferred extremal correspond to the non-determinism of p-adic differential equations due to the presence of pseudo constants which have vanishing p-adic derivative. Pseudo- constants are functions depend on finite number of pinary digits of its arguments.
2. The quantum criticality of TGD is suggested to be realized in in terms of infinite hierarchies of super-symplectic symmetry breakings in the sense that only a sub-algebra with conformal weights which are n-multiples of those for the entire algebra act as conformal gauge symmetries. This might be true for all conformal algebras involved. One has fractal hierarchy since the sub-algebras in question are isomorphic: only the scale of conformal gauge symmetry increases in the phase transition increasing n. The hierarchies correspond to sequences of integers $n(i)$ such tht $n(i)$ divides $n(i+1)$. These hierarchies would very naturally correspond to hierarchies of inclusions of hyper-finite factors and $m(i)= n(i+1)/n(i)$ could correspond to the integer n characterizing the index of inclusion, which has value $n\geq$ 3. Possible problem is that $m(i)=2$ would not correspond to Jones inclusion. Why the

scaling by power of two would be different? The natural question is whether the primes dividing n(i) or m(i) could define the preferred primes.

3. Negentropic entanglement corresponds to entanglement for which density matrix is projector. For n-dimensional projector any prime p dividing n gives rise to negentropic entanglement in the sense that the number theoretic entanglement entropy defined by Shannon formula by replacing p_i in $\log(p_i)= \log(1/n)$ by its p-adic norm $N_p(1/n)$ is negative if p divides n and maximal for the prime for which the dividing power of prime is largest power-of-prime factor of n. The identification of p-adic primes as factors of n is highly attractive idea. The obvious question is whether n corresponds to the integer characterizing a level in the hierarchy of conformal symmetry breakings.

4. The adelic picture about TGD led to the question whether the notion of unitary could be generalized. S-matrix would be unitary in adelic sense in the sense that $P_m=(SS^\dagger)_{mm}=1$ would generalize to adelic context so that one would have product of real norm and p-adic norms of P_m. In the intersection of the realities and p-adicities P_m for reals would be rational and if real and p-adic P_m correspond to the same rational, the condition would be satisfied. The condition that $P_m\leq 1$ seems however natural and forces separate unitary in each sector so that this options seems too tricky.

These are the basic ideas that I have discussed hitherto.

2. Could preferred primes characterize algebraic extensions of rationals?

The intuitive feeling is that the notion of preferred prime is something extremely deep and the deepest thing I know is number theory. Does one end up with preferred primes in number theory? This question brought to my mind the notion of ramification of primes (see this) (more precisely, of prime ideals of number field in its extension), which happens only for special primes in a given extension of number field, say rationals. Could this be the mechanism assigning preferred prime(s) to a given elementary system, such as elementary particle? I have not considered their role earlier also their hierarchy is highly relevant in the number theoretical vision about TGD.

1. Stating it very roughly (I hope that mathematicians tolerate this language): As one goes from number field K, say rationals Q, to its algebraic extension L, the original prime ideals in the so called integral closure (see this) over integers of K decompose to products of prime ideals of L (prime is a more rigorous manner to express primeness).

    Integral closure for integers of number field K is defined as the set of elements of K, which are roots of some monic polynomial with coefficients, which are integers of K and having the form $x^n+a_{n-1}x^{n-1}+...+a_0$. The integral closures of both K and L are considered. For instance, integral closure of algebraic extension of K over K is the extension itself. The integral closure of complex numbers over ordinary integers is the set of algebraic numbers.

2. There are two further basic notions related to ramification and characterizing it. Relative discriminant is the ideal divided by all ramified ideals in K and relative different is the ideal of L divided by all ramified P_i:s. Note that te general ideal is analog of integer and these ideas represent the analogous of product of preferred primes P of K and primes P_i of L dividing them.

3. A physical analogy is provided by decomposition of hadrons to valence quarks. Elementary particles becomes composite of more elementary particles in the extension. The decomposition to these more elementary primes is of form $P= \prod P_i^{e(i)}$, where e_i is the ramification index - the physical analog would be the number of elementary particles of type i in the state (see this). Could the ramified rational primes could define the physically preferred primes for a given elementary system?

In TGD framework, the extensions of rationals (see this) and p-adic number fields (see this) are unavoidable and interpreted as an evolutionary hierarchy physically and cosmological evolution would have gradually proceeded to more and more complex extensions. One can say that string world sheets and partonic 2-surfaces with parameters of defining functions in increasingly complex extensions of prime emerge during evolution. Therefore ramifications and the preferred primes defined by them are unavoidable. For p-adic number fields the number of extensions is much smaller for instance for p>2 there are only 3 quadratic extensions.

1.

2. In p-adic context a proper definition of counterparts of angle variables as phases allowing definition of the analogs of trigonometric functions requires the introduction of algebraic extension giving rise to some roots of unity. Their number depends on the angular resolution. These roots allow to define the counterparts of ordinary trigonometric functions (the naive generalization based on Taylor series is not periodic) and also allows to defined the counterpart of definite integral in these degrees of freedom as discrete Fourier analysis. For the simplest algebraic extensions defined by x^n-1 for which Galois group is abelian are are unramified so that something else is needed. One has decomposition $P= \prod P_i^{e(i)}$, $e(i)=1$, analogous to n-fermion state so that simplest cyclic extension does not give rise to a ramification and there are no preferred primes.

3. What kind of polynomials could define preferred algebraic extensions of rationals? Irreducible polynomials are certainly an attractive candidate since any polynomial reduces to a product of them. One can say that they define the elementary particles of number theory. Irreducible polynomials have integer coefficients having the property that they do not decompose to products of polynomials with rational coefficients. It would be wrong to say that only these algebraic extensions can appear but there is a temptation to say that one can reduce the study of extensions to their study. One can even consider the possibility that string world sheets associated with products of irreducible polynomials are unstable against decay to those characterize irreducible polynomials.

4. What can one say about irreducible polynomials? Eisenstein criterion states following. If $Q(x)= \sum_{k=0,...,n} a_k x^k$ is n:th order polynomial with integer coefficients and with the property that there exists at least one prime dividing all coefficients a_i except a_n and that p^2 does not divide a_0, then Q is irreducible. Thus one can assign one or more preferred primes to the algebraic extension defined by an irreducible polynomial Q - in fact any polynomial allowing ramification. There are also other

kinds of irreducible polynomials since Eisenstein's condition is only sufficient but not necessary.

5. Furthermore, in the algebraic extension defined by Q, the primes P having the above mentioned characteristic property decompose to an n :th power of single prime P_i: $P = P_i^n$. The primes are maximally/completely ramified. The physical analog $P = P_0^n$ is Bose-Einstein condensate of n bosons. There is a strong temptation to identify the preferred primes of irreducible polynomials as preferred p-adic primes.

A good illustration is provided by equations $x^2 + 1 = 0$ allowing roots $x_{+/-} = +/- i$ and equation $x^2 + 2px + p = 0$ allowing roots $x_{+/-} = -p +/- p^{1/2}p - 1^{1/2}$. In the first case the ideals associated with $+/- i$ are different. In the second case these ideals are one and the same since $x_+ = =- x_- + p$: hence one indeed has ramification. Note that the first example represents also an example of irreducible polynomial, which does not satisfy Eisenstein criterion. In more general case the n conditions on defined by symmetric functions of roots imply that the ideals are one and same when Eisenstein conditions are satisfied.

6. What does this mean in p-adic context? The identity of the ideals can be stated by saying $P = P_0^n$ for the ideals defined by the primes satisfying the Eisenstein condition. Very loosely one can say that the algebraic extension defined by the root involves n^{th} root of p-adic prime p. This does not work! Extension would have a number whose n^{th} power is zero modulo p. On the other hand, the p-adic numbers of the extension modulo p should be finite field but this would not be field anymore since there would exist a number whose n^{th} power vanishes. The algebraic extension simply does not exist for preferred primes. The physical meaning of this will be considered later.

7. What is so nice that one could readily construct polynomials giving rise to given preferred primes. The complex roots of these polymials could correspond to the points of partonic 2-surfaces carrying fermions and defining the ends of boundaries of string world sheet. It must be however emphasized that the form of the polynomial depends on the choices of the complex coordinate. For instance, the shift $x \rightarrow x+1$ transforms $(x^n-1)/(x-1)$ to a polynomial satisfying the Eisenstein criterion. One should be able to fix allowed coordinate changes in such a manner that the extension remains irreducible for all allowed coordinate changes.

Already the integral shift of the complex coordinate affects the situation. It would seem that only the action of the allowed coordinate changes must reduce to the action of Galois group permuting the roots of polynomials. A natural assumption is that the complex coordinate corresponds to a complex coordinate transforming linearly under subgroup of isometries of the imbedding space.

In the general situation one has $P = \prod P_i^{e(i)}$, $e(i) \geq 1$ so that aso now there are prefered primes so that the appearance of preferred primes is completely general phenomenon.

3. A connection with Langlands program?

In Langlands program (see this) the great vision is that the n-dimensional representations of Galois groups G characterizing algebraic extensions of rationals or more general number fields define n-dimensional adelic representations of adelic Lie groups, in particular the adelic linear group Gl(n,A). This would mean that it is possible to reduce these representations to a number theory for adeles. This would be highly relevant in the vision about TGD as a generalized number theory. I have speculated with this possibility earlier (see this) but the mathematics is so horribly abstract that it takes decade before one can have even hope of building a rough vision.

One can wonder whether the irreducible polynomials could define the preferred extensions K of rationals such that the maximal abelian extensions of the fields K would in turn define the adeles utilized in Langlands program. At least one might hope that everything reduces to the maximally ramified extensions.

At the level of TGD string world sheets with parameters in an extension defined by an irreducible polynomial would define an adele containing various p-adic number fields defined by the primes of the extension. This would define a hierarchy in which the prime ideals of previous level would decompose to those of the higher level. Each irreducible extension of rationals would correspond to some physically preferred p-adic primes.

It should be possible to tell what the preferred character means in terms of the adelic representations. What happens for these representations of Galois group in this case? This is known.

1. For Galois extensions ramification indices are constant: e(i)=e and Galois group acts transitively on ideals P_i dividing P. One obtains an n-dimensional representation of Galois group. Same applies to the subgroup of Galois group G/I where I is subgroup of G leaving P_i invariant. This group is called inertia group. For the maximally ramified case G maps the ideal P_0 in $P=P_0^n$ to itself so that G=I and the action of Galois group is trivial taking P_0 to itself, and one obtains singlet representations.
2. The trivial action of Galois group looks like a technical problem for Langlands program and also for TGD unless the singletness of P_i under G has some physical interpretation. One possibility is that Galois group acts as like a gauge group and here the hierarchy of sub-algebras of super-symplectic algebra labelled by integers n is highly suggestive. This raises obvious questions. Could the integer n characterizing the sub-algebra of super-symplectic algebra acting as conformal gauge transformations, define the integer defined by the product of ramified primes? P_0^n brings in mind the n conformal equivalence classes which remain invariant under the conformal transformations acting as gauge transformiations. . Recalling that relative discriminant is an of K ideal divisible by ramified prime ideals of K, this means that n would correspond to the relative discriminant for K=Q.

Are the preferred primes those which are "physical" in the sense that one can assign to the states satisfying conformal gauge conditions?

4. A connection with infinite primes?

Infinite primes are one of the mathematical outcomes of TGD. There are two kinds of infinite primes. There are the analogs of free many particle states consisting of fermions and bosons labelled by primes of the previous level in the hierarchy. They correspond to states of a supersymmetric arithmetic quantum field theory or actually a hierarchy of them obtained by a repeated second quantization of this theory. A connection between infinite primes representing bound states and and irreducible polynomials is highly suggestive.

1. The infinite prime representing free many-particle state decomposes to a sum of infinite part and finite part having no common finite prime divisors so that prime is obtained. The infinite part is obtained from "fermionic vacuum" $X = \prod_k p_k$ by dividing away some fermionic primes p_i and adding their product so that one has $X \to X/m + m$, where m is square free integer. Also m=1 is allowed and is analogous to fermionic vacuum interpreted as Dirac sea without holes. X is infinite prime and pure many-fermion state physically. One can add bosons by multiplying X with any integers having no common denominators with m and its prime decomposition defines the bosonic contents of the state. One can also multiply m by any integers whose prime factors are prime factors of m.
2. There are also infinite primes, which are analogs of bound states and at the lowest level of the hierarchy they correspond to irreducible polynomials P(x) with integer coefficients. At the second levels the bound states would naturally correspond to irreducible polynomials $P_n(x)$ with coefficients $Q_k(y)$, which are infinite integers at the previous level of the hierarchy.
3. What is remarkable that bound state infinite primes at given level of hierarchy would define maximally ramified algebraic extensions at previous level. One indeed has infinite hierarchy of infinite primes since the infinite primes at given level are infinite primes in the sense that they are not divisible by the primes of the previous level. The formal construction works as such. Infinite primes correspond to polynomials of single variable at the first level, polynomials of two variables at second level, and so on. Could the Langlands program could be generalized from the extensions of rationals to polynomials of complex argument and that one would obtain infinite hierarchy?
4. Infinite integers in turn could correspond to products of irreducible polynomials defining more general extensions. This raises the conjecture that infinite primes for an extension K of rationals could code for the algebraic extensions of K quite generally. If infinite primes correspond to real quantum states they would thus correspond the extensions of rationals to which the parameters appearing in the functions defining partonic 2-surfaces and string world sheets.
   This would support the view that partonic 2-surfaces associated with algebraic extensions defined by infinite integers and thus not irreducible are unstable against decay to partonic 2-surfaces which corresponds to extensions assignable to infinite primes. Infinite composite integer defining intermediate unstable state would decay to its composites. Basic particle physics phenomenology would have number theoretic analog and even more.

5. According to Wikipedia, Eisenstein's criterion (see this) allows generalization and what comes in mind is that it applies in exactly the same form also at the higher levels of the hierarchy. Primes would be only replaced with prime polynomials and the there would be at least one prime polynomial Q(y) dividing the coefficients of $P_n(x)$ except the highest one such that its square would not divide P_0. Infinite primes would give rise to an infinite hierarchy of functions of many complex variables. At first level zeros of function would give discrete points at partonic 2-surface. At second level one would obtain 2-D surface: partonic 2-surfaces or string world sheet. At the next level one would obtain 4-D surfaces. What about higher levels? Does one obtain higher dimensional objects or something else. The union of n 2-surfaces can be interpreted also as 2n-dimensional surface and one could think that the hierarchy describes a hierarchy of unions of correlated partonic 2-surfaces. The correlation would be due to the preferred extremal property of Kähler action.

One can ask whether this hierarchy could allow to generalize number theoretical Langlands to the case of function fields using the notion of prime function assignable to infinite prime. What this hierarchy of polynomials of arbitrary many complex arguments means physically is unclear. Do these polynomials describe many-particle states consisting of partonic 2-surface such that there is a correlation between them as sub-manifolds of the same space-time sheet representing a preferred extremals of Kähler action?

This would suggest strongly the generalization of the notion of p-adicity so that it applies to infinite primes.

1. This looks sensible and maybe even practical! Infinite primes can be mapped to prime polynomials so that the generalized p-adic numbers would be power series in prime polynomial - Taylor expansion in the coordinate variable defined by the infinite prime. Note that infinite primes (irreducible polynomials) would give rise to a hierarchy of preferred coordinate variables. In terms of infinite primes this expansion would require that coefficients are smaller than the infinite prime P used. Are the coefficients lower level primes? Or also infinite integers at the same level smaller than the infinite prime in question? This criterion makes sense since one can calculate the ratios of infinite primes as real numbers.
2. I would guess that the definition of infinite-P p-adicity is not a problem since mathematicians have generalized the number theoretical notions to such a level of abstraction much above of a layman like me. The basic question is how to define p-adic norm for the infinite primes (infinite only in real sense, p-adically they have unit norm for all lower level primes) so that it is finite.
3. There exists an extremely general definition of generalized p-adic number fields (see this). One considers Dedekind domain D, which is a generalization of integers for ordinary number field having the property that ideals factorize uniquely to prime ideals. Now D would contain infinite integers. One introduces the field E of fractions consisting of infinite rationals.
Consider element e of E and a general fractional ideal eD as counterpart of ordinary rational and decompose it to a ratio of products of powers of ideals defined by prime ideals, now those defined by infinite primes. The general expression for the p-

adic norm of x is $x^{-ord(P)}$, where n defines the total number of ideals P appearing in the factorization of a fractional ideal in E: this number can be also negative for rationals. When the residue field is finite (finite field G(p,1) for p-adic numbers), one can take c to the number of its elements (c=p for p-adic numbers.

Now it seems that this number is not finite since the number of ordinary primes smaller than P is infinite! But this is not a problem since the topology for completion does not depend on the value of c. The simple infinite primes at the first level (free many-particle states) can be mapped to ordinary rationals and q-adic norm suggests itself: could it be that infinite-P p-adicity corresponds to q-adicity discussed by Khrennikov about p-adic analysis. Note however that q-adic numbers are not a field.

Finally, a loosely related question. Could the transition from infinite primes of K to those of L takes place just by replacing the finite primes appearing in infinite prime with the decompositions? The resulting entity is infinite prime if the finite and infinite part contain no common prime divisors in L. This is not the case generally if one can have primes P_1 and P_2 of K having common divisors as primes of L: in this case one can include P_1 to the infinite part of infinite prime and P_2 to finite part.

For details see the article The Origin of Preferred p-Adic Primes? For a summary of earlier postings see Links to the latest progress in TGD.

posted by Matti Pitkanen @ 2:34 AM

04/27/2015   -   http://matpitka.blogspot.com/2015/04/could-adelic-approach-allow-to.html#comments

Could adelic approach allow to understand the origin of preferred p-adic primes?

The comment of Crow to the posting Intentions, cognitions, and time stimulated rather interesting ideas about the adelization of Quantum-TGD.

First two questions.

1. What is Adelic Quantum-TGD? The basic vision is that scattering amplitudes are obtained by algebraic continuation to various number fields from the intersection of realities and p-adicities (briefly *intersection* in what follows) represented at the space-time level by string world sheets and partonic 2-surfaces for which defining parameters (WCW coordinates) are in rational or in in some algebraic extension of p-adic numbers. This principle is a combination of strong form of holography and algebraic continuation as a manner to achieve number theoretic universality.

2. Why Adelic quantum TGD? Adelic approach is free of the earlier assumptions, which require mathematics which need not exist: transformation of p-adic space-time surfaces to real ones as a realization of intentional actions was the questionable assumption, which is un-necessary if cognition and matter are two different aspects of existence as already the success of p-adic mass calculations strongly suggests. It always takes years to develop ability to see things from bigger perspective and distill discoveries from clever inventions. Now adelicity is totally obvious. Being a conservative radical - not radical radical or radical conservative - is the correct strategy which I have been gradually learning. This particular lesson was excellent!

Some years ago, Crow sent to me the book of Lapidus about adelic strings. Witten wrote for long time ago an article in which the demonstrated that that the product of real stringy vacuum amplitude and its p-adic variants equals to 1. This is a generalisation of the adelic identity for a rational number stating that the product of the norm of rational number with its p-adic norms equals to one.

The real amplitude in the intersection of realities and p-adicities for all values of parameter is rational number or in an appropriate algebraic extension of rationals. If given p-adic amplitude is just the p-adic norm of real amplitude, one would have the adelic identity. This would however require that p-adic variant of the amplitude is real number-valued: I want p-adic valued amplitudes. A further restriction is that Witten's adelic identity holds for vacuum amplitude. I live in Zero Energy Ontology (ZEO) and want it for entire S-matrix, M-matrix, and/or U-matrix and for all states of the basis in some sense.

Consider first the vacuum amplitude. A weaker form of the identity would be that the *p-adic norm* of a given p-adic valued amplitude is same as that p-adic norm for the rational-valued real amplitude (this generalizes to algebraic extensions, I dare to guess) in the intersection. This would make sense and give a non-trivial constraint: algebraic continuation would guarantee this constraint. In particular, the p-adic norm of the real amplitude would be inverse of the product of p-adic norms of p-adic amplitudes. Most of these amplitudes should have p-adic norm equal to one in other words, real amplitude is product of finite number of powers of prime. This because the p-adic norms must approach rapidly to unity as p-adic prime increases and for large p-adic primes this means that the norm is exactly unity. Hence the p-adic norm of p-adic amplitude equals to 1 for most primes.

In ZEO one must consider S-, M-, or U-matrix elements. U and S are unitary. M is product of hermitian square root of density matrix times unitary S-matrix. Consider next S-matrix.

1. For S-matrix elements one should have $p_m=(SS^{\dagger})_{mm}=1$. This states the unitarity of S-matrix. Probability is conserved. Could it make sense to generalize this condition and demand that it holds true only adelically that only for the product of real and p-adic norms of p_m equals to one: $N_R(p_m)(R)\prod_p N_p(p_m(p))=1$. This could be actually true identically in the intersection if algebraic continuation principle holds true. Despite the triviality of the adelicity condition, one need not have anymore unitarity separately for reals and p-adic number fields. Notice that the numbers p_m would be arbitrary rationals in the most general cased.
2. Could one even replace N_p with canonical identification or some form of it with cutoffs reflecting the length scale cutoffs? Canonical identification behaves for powers of p like p-adic norm and means only more precise map of p-adics to reals.
3. For a given diagonal element of unit matrix characterizing particular state m one would have a product of real norm and p-adic norms. The number of the norms, which differ from unity would be finite. This condition would give finite number of exceptional p-adic primes, that is assign to a given quantum state m a *finite number of preferred p-adic primes*! I have been searching for a long time the underlying deep reason for this assignment forced by the p-adic mass calculations and here it might be.
4. Unitarity might thus fail in real sector and in a finite number of p-adic sectors (otherwise the product of p-adic norms would be infinite or zero). In some sense the failures would compensate each other in the adelic picture. The failure of course brings in mind p-adic thermodynamics, which indeed means that adelic $SS^{\dagger}$, or should it be called $MM^{\dagger}$, is not unitary but defines the density matrix defining the p-adic thermal state! Recall that M-matrix is defined as hermitian square root of density matrix and unitary S-matrix.
5. The weakness of these arguments is that states are assumed to be labelled by discrete indices. Finite measurement resolution implies discretization and could justify this.

The p-adic norms of p_m or the images of p_m under canonical identification in a given number field would define analogs of probabilities. Could one indeed have $\sum_m p_m=1$ so that $SS^{\dagger}$ would define a density matrix?

1. For the ordinary S-matrix this cannot be the case since the sum of the probabilities p_m equals to the dimension N of the state space: $\sum p_m=N$. In this case one could accept $p_m>1$ both in real and p-adic sectors. For this option adelic unitarity would make sense and would be highly non-trivial condition allowing perhaps to understand how preferred p-adic primes emerge at the fundamental level.
2. If S-matrix is multiplied by a hermitian square root of density matrix to get M-matrix, the situation changes and one indeed obtains $\sum p_m=1$. $MM^{\dagger}=1$ does not make sense anymore and must be replaced with $MM^{\dagger}=\rho$, in special case a projector to a N-dimensional subspace proportional to $1/N$. In this case the numbers $p(m)$ would have p-adic norm larger than one for the divisors of N and would define preferred p-adic primes. For these primes the sum $N_p(p(m))$ would not be equal to 1 but to $NN_p(1/N)$.

3. Situation is different for hyper-finite factors of type II$_1$ for which the trace of unit matrix equals to one by definition and MM†=1 and $\sum$ p$_m$=1 with sum defined appropriately could make sense. If MM† could be also a projector to an infinite-D subspace. Could the M-matrix using the ordinary definition of dimension of Hilbert space be equivalent with S-matrix for the state space using the definition of dimension assignable to HFFs? Could these notions be dual of each other? Could the adelic S-matrix define the counterpart of M-matrix for HFFs?

This looks like a nice idea but usually good looking ideas do not live long in the crossfire of counter arguments. The following is my own. The reader is encouraged to invent his or her own objections.

1. The most obvious objection against the very attractive *direct* algebraic continuation} from real to p-adic sector is that if the real norm or real amplitude is small then the p-adic norm of its p-adic counterpart is large so that p-adic variants of p$_m$(p) can become larger than 1 so that probability interpretation fails. As noticed there is no actually no need to pose probability interpretation. The only way to overcome the "problem" is to assume that unitarity holds separately in each sector so that one would have p(m)=1 in all number fields but this would lead to the loss of preferred primes.
2. Should p-adic variants of the real amplitude be defined by canonical identification or its variant with cutoffs? This is mildly suggested by p-adic thermodynamics. In this case it might be possible to satisfy the condition p$_m$(R)$\prod_p$ N$_p$(p$_m$(p))=1. One can however argue that the adelic condition is an ad hoc condition in this case.

To sum up, if the above idea survives all the objections, it could give rise to a considerable progress. A first principle understanding of how preferred p-adic primes are assigned to quantum states and thus a first principle justification for p-adic thermodynamics. For the ordinary definition of S-matrix this picture makes sense and also for M-matrix. One would still need the justification of canonical identification map playing a key role in p-adic thermodynamics allowing to map p-adic mass squared to its real counterpart.

posted by Matti Pitkanen @ <u>1:16 AM</u>

04/26/2015 - <u>http://matpitka.blogspot.com/2015/04/hierarchies-of-conformal-symmetry.html#comments</u>

Hierarchies of conformal symmetry breakings, quantum criticalities, Planck constants, and hyper-finite factors

TGD is characterized by various hierarchies. There are fractal hierarchies of quantum criticalities, Planck constants and hyper-finite factors and these hierarchies relate to hierarchies of space-time sheets, and selves. These hierarchies are closely related and this article describes these connections. In this article the recent view about connections between various hierarchies associated with quantum TGD are described.

For details, see the article Hierarchies of conformal symmetry breakings, quantum criticalities, Planck constants, and hyper-finite factors. For a summary of earlier postings see Links to the latest progress in TGD.

04/26/2015 - http://matpitka.blogspot.com/2015/04/updated-view-about-k-geometry-of-wcw.html#comments

Updated View about Kähler geometry of WCW

Quantum-TGD reduces to a construction of Kähler geometry for what I call the "World of Classical Worlds. It has been clear from the beginning that the gigantic super-conformal symmetries generalizing ordinary super-conformal symmetries are crucial for the existence of WCW Kähler metric. The detailed identification of Kähler function and WCW Kähler metric has however turned out to be a difficult problem.

It is now clear that WCW geometry can be understood in terms of the analog of AdS/CFT duality between fermionic and space-time degrees of freedom (or between Minkowskian and Euclidian space-time regions) allowing to express Kähler metric either in terms of Kähler function or in terms of anti-commutators of WCW gamma matrices identifiable as super-conformal Noether super-charges for the symplectic algebra assignable to $\delta M^4_{+/-} \times CP_2$. The string model type description of gravitation emerges and also the TGD based view about dark matter becomes more precise. String tension is however dynamical rather than pregiven and the hierarchy of Planck constants is necessary in order to understand the formation of gravitationally bound states. Also the proposal that sparticles correspond to dark matter becomes much stronger: sparticles actually are dark variants of particles.

A crucial element of the construction is the assumption that super-symplectic and other super-conformal symmetries having the same structure as 2-D super-conformal groups can be seen a broken gauge symmetries such that sub-algebra with conformal weights coming as n-ples of those for full algebra act as gauge symmetries. In particular, the Noether charges of this algebra vanish for preferred extremals- this would realize the strong form of holography implied by strong form of General Coordinate Invariance.

This gives rise to an infinite number of hierarchies of conformal gauge symmetry breakings with levels labelled by integers $n(i)$ such that $n(i)$ divides $n(i+1)$ interpreted as hierarchies of dark matter with levels labelled by the value of Planck constant $h_{eff}=n\times h$. These hierarchies define also hierarchies of quantum criticalities and are proposed to give rise to inclusion hierarchies of hyperfinite factors of II_1 having interpretation in terms of finite cognitive resolution. These hierarchies would be fundamental for the understanding of living matter.

For details see the article Updated view about Kähler geometry of WCW. For a summary of earlier postings see Links to the latest progress in TGD.

04/26/2015   -   http://matpitka.blogspot.com/2015/04/intentions-cognition-and-time.html#comments

Intentions, Cognition, and Time

Intentions involve time in an essential manner and this led to the idea that p-adic-to-real quantum jumps could correspond to a realization of intentions as actions. It however seems that this hypothesis posing strong additional mathematical challenges is not needed if one accepts adelic approach in which real space-time time and its p-adic variants are all present and quantum physics is adelic. I have already earlier developed the first formulation of p-adic space-time surfaces as cognitive charges of real space-time surfaces and also the ideas related to the adelic vision.

The recent view involving strong form of holography would provide dramatically simplified view about how these representations are formed as continuations of representations of strings world sheets and partonic 2-surfaces in the intersection of real and p-adic variants of WCW ("World of Classical Worlds") in the sense that the parameters characterizing these representations are in the algebraic numbers in the algebraic extension of p-adic numbers involved.

For details see the article [Intentions, Cognition, and Time](). For a summary of earlier postings see [Links to the latest progress in TGD]().

posted by Matti Pitkanen @ 12:14 AM   4 comments

4 Comments:

At 12:29 PM, Anonymous said...

Matti, I really think you are on to something (as if you weren't already) with the adelic approach to things. I was just looking at a paper in wrote on fractal strings and membranes a few years ago and realized that the adelic product defined by Lapidus and others, assigns to each element of the product a (square intehrable) Hilbert space.

Also I think there must be something special about modular arithmetic , the special status of the integral numbers 12 and 24 hours there, and the approximately 24hr period of the standard day. Maybe life requires these almost ratios? If other planets had cycles not commensurate then perhaps life would not be favored? --crow

At 10:09 PM, Matpitka@luukku.com said...

To Anonymous:

Thank you for a very stimulating comment.

Adelic approach is free of assumptions which require mathematics which need not exist: transformation of p-adic space-time surfaces to real ones was the questionable assumption. It always takes years to develop ability to see it from bigger perspective. Now adelicity is totally obvious. Being a conservative radical - not radical radical or radical conservative - is the correct strategy which I have been gradually learning. An excellent lesson!

I think you sent the book of Lapidus about adelic strings. Witten wrote for long time ago an article in which the demonstrated that that the product of real stringy vacuum amplitude and its p-adic variants equals to 1. This is a generalisation of the adelic identity for a rational number.

One can think that the real amplitude in the intersection of realities and p-adicities for all values of parameter is rational number. If given p-adic amplitude is just the p-adic norm of real amplitude, one would have the adelic identity. But this would require that p-adic variant of the amplitude is real number-valued: I have p-adic valued amplitudes. And Witten's identity holds for vacuum amplitude. I want it for entire S-matrix, M-matrix, and/or U-matrix and for all states of the basis in some sense.

a) Consider first vacuum amplitude. A weaker form of the identity would be that the *p-adic norm* of given p-adic valued amplitude is same as that p-adic norm for the rational-valued real amplitude (this generalizes to algebraic extensions, I dare to guess). This would make sense and give a non-trivial constraint. In particular, the p-adic norm of the real amplitude would be inverse of the the product of p-adic norms of p-adic amplitudes. Most of these amplitudes should have p-adic norm equal to one. This condition can make sense only if the p-adic norm of p-adic amplitude equals to 1 for most prime.

b) In ZEO one must consider S-, M-, or U-matrix elements. U and S are unitary. M is product of hermitian square root of density matrix times unitary S-matrix. Consider the S-matrix.

*For S-matrix elements one should have SS^dagger=1. This states unitarity of S-matrix. Probability is conserved. Could it make sense to generalize this condition and demand that it holds true only adelically that is for the product of real and p-adic norms of SS^dagger in various number fields? For each state m of basis: the product of norm real X_mm =(SS^dagger)_{mm} and p-adic norms its p-adic counterparts would be qual to 1 in the intersection of reality and p-adicities. Strong condition would be that this holds for X_mm themselves. It could if the the p-adic norm of X_mm is X_mm itself that is power of p.

*For a given diagonal element of unit matrix defining particular state m one would have a product of real norm and p-adic norms. The number of the norms, which differ from unity would be finite. This condition would give finite number of exceptional p-adic primes, that is assign to a given quantum state labelling the diagonal matrix element of SS^dagger a *finite number of preferred p-adic primes*!! The underlying deep reason for this assignment I have been looking for!!

*Unitary might thus fail in real sector and in a finite number of p-adic sectors (otherwise the product of p-adic norms would be infinite or zero). In some sense the failures would compensate each other in the adelic picture. The failure of course brings in mind p-adic thermo-dynamics which indeed means that SS^+ is density matrix defining the p-adic thermal state!

*The diagonal elements of $SS^{\dagger}_{\{mm\}}$ in given number field would define analogs of probabilities. Could these probabilities be interpreted as the probabilities, whose sum equals to 1? Probability conservation for a given number field. Adelic S-matrix would be the more sophisticated counterpart of M-matrix!

One can consider a variant of this. One could also consider the possibility that the p-adic norm of $SS^{\dagger}\_{mm}$ is replaced with its image under canonical identification. The information loss wold not be so huge. This might be required by p-adic thermodynamics.

At <u>10:14 PM</u>, <u>Matpitka@luukku.com</u> said...

To Anonymous:

I forgot the summary. The outcome could be two breakthroughs. First principle understanding of how preferred p-adic primes are assigned to quantum states and first principle justification for p-adic thermodynamics.

04/25/2015 - http://matpitka.blogspot.com/2015/04/good-and-evil-life-and-death.html#comments

Good and Evil, Life and Death

In principle the proposed conceptual framework allows already now a consideration of the basic questions relating to concepts like Good and Evil and Life and Death. Of course, too many uncertainties are involved to allow any definite conclusions, and one could also regard the speculations as outputs of the babbling period necessarily accompanying the development of the linguistic and conceptual apparatus making ultimately possible to discuss these questions more seriously.

Even the most hard-boiled materialistic sceptic mentions ethics and moral when suffering personal injustice. Is there actual justification for moral laws? Are they only social conventions or is there some hard core involved? Is there some basic ethical principle telling what deeds are good and what deeds are bad?

Second group of questions relates to life and biological death. How should on define life? What happens in the biological death? Is self preserved in the biological death in some form? Is there something deserving to be called soul? Are reincarnations possible? Are we perhaps responsible for our deeds even after our biological death? Could the law of Karma be consistent with physics? Is liberation from the cycle of Karma possible?

In the sequel, these questions are discussed from the point of view of TGD inspired theory of consciousness. It must be emphasized that the discussion represents various points of view rather than being a final summary. Also mutually conflicting points of view are considered. The cosmology of consciousness, the concept of self having space-time sheet and causal diamond as its correlates, the vision about the fundamental role of negentropic entanglement, and the hierarchy of Planck constants identified as hierarchy of dark matters and of quantum critical systems, provide the building blocks needed to make guesses about what biological death could mean from subjective point of view.

For details see the article Good and Evil, Life and Death. For a summary of earlier postings see Links to the latest progress in TGD.

posted by Matti Pitkanen @ 10:23 PM

2 Comments:

At 4:47 AM, Anonymous said...

The Zen-koan about 500-years as a fox is about karma and denial of. Liberation from karma is not denial of karma, nor rule of karma - the opposite of liberation. I'm tempted to formulate in this context, that even in liberated state karma does not cease being observable, even though karma is not being fed causal metabolic energy.

And strangely, "saint" does not become saint by excluding but by including every "sinner". "Saint" that morally-judgmentally excludes sinners from "higher-self" is the worst "sinner", the worst Jungian shadow-projector. Those who do most horrible, most hurting deeds are not the self-loving and self-pleasing "sinners" but the White Knights of Light and Truth, who build highway to hell paved with "good intentions".

At 5:50 AM, Matpitka@luukku.com said...

Exactly. Good deeds generate negentropic entanglement with owner sinners and is gift to them too.

04/24/2015 - http://matpitka.blogspot.com/2015/04/variation-of-newstons-constant-and-of.html#comments

Variation of Newton's constant and of length of day

J. D. Anderson et al have published an article discussing the observations suggesting a periodic variation of the measured value of Newton constant and variation of length of day (LOD) (see also this). This article represents TGD based explanation of the observations in terms of a variation of Earth radius. The variation would be due to the pulsations of Earth coupling via gravitational interaction to a dark matter shell with mass about $1.3\times 10^{-4}M_E$ introduced to explain Flyby anomaly: the model would predict $\Delta G/G = 2\Delta R/R$ and $\Delta LOD/LOD = 2\Delta R_E/R_E$ with the variations pf G and length of day in opposite phases. The expermental finding $\Delta R_E/R_E = M_D/M_E$ is natural in this framework but should be deduced from first principles.

The gravitational coupling would be in radial scaling degree of freedom and rigid body rotational degrees of freedom. In rotational degrees of freedom the model is in the lowest order approximation mathematically equivalent with Kepler model. The model for the formation of planets around Sun suggests that the dark matter shell has radius equal to that of Moon's orbit. This leads to a prediction for the oscillation period of Earth radius: the prediction is consistent with the observed 5.9 years period. The dark matter shell would correspond to n=1 Bohr orbit in the earlier model for quantum gravitational bound states based on large value of Planck constant. Also n>1 orbits are suggestive and their existence would provide additional support for TGD view about quantum gravitation.

For details see the chapter Cosmology and Astrophysics in Many-Sheeted Space-Time or the article Variation of Newton's constant and of length of day. For a summary of earlier postings see Links to the latest progress in TGD.

04/21/2015 - http://matpitka.blogspot.com/2015/04/connection-between-boolean-cognition.html#comments

Connection between Boolean cognition and emotions

Weak form of NMP allows the state function reduction to occur in 2^n-1 manners corresponding to subspaces of the sub-space defined by n-dimensional projector if the density matrix is n-dimensional projector (the outcome corresponding to 0-dimensional subspace and is excluded). If the probability for the outcome of state function reduction is same for all values of the dimension $1 \leq m \leq n$, the probability distribution for outcome is given by binomial distribution $B(n,p)$ for $p=1/2$ (head and tail are equally probable) and given by $p(m)= b(n,m) \times 2^{-n}= (n!/m!(n-m)!) \times 2^{-n}$. This gives for the average dimesion $E(m)= n/2$ so that the negentropy would increase on the average. The world would become gradually better.

One cannot avoid the idea that these different degrees of negentropic entanglement could actually give a realization of Boolean algebra in terms of conscious experiences.

1. Could one speak about a hierarchies of codes of cognition based on the assignment of different degrees of "feeling good" to the Boolean statements? If one assumes that the nth bit is always 1, all independent statements except one correspond at least two non-vanishing bits and corresponds to negentropic entanglement. Only of statement (only last bit equal to 1) would correspond 1 bit and to state function reduction reducing the entanglement completely (brings in mind the fruit in the tree of Good and Bad Knowlege!).
2. A given hierarchy of breakings of super-symplectic symmetry corresponds to a hierarchy of integers $n_{i+1}= \prod_{k \leq i} m_k$. The codons of the first code would consist of sequences of m_1 bits. The codons of the second code consists of m_2 codons of the

first code and so on. One would have a hierarchy in which codons of previous level become the letters of the code words at the next level of the hierarchy.

In fact, I ended up with almost Boolean algebra for decades ago when considering the hierarchy of genetic codes suggested by the hierarchy of Mersenne primes $M(n+1)= M_{M(n)}$, $M_n= 2^n-1$.

1. The hierarchy starting from $M_2=3$ contains the Mersenne primes $3,7,127,2^{127}-1$ and Hilbert conjectured that all these integers are primes. These numbers are almost dimensions of Boolean algebras with n=2,3,7,127 bits. The maximal Boolean sub-algebras have m=n-1=1,2,6,126 bits.
2. The observation that m=6 gives 64 elements led to the proposal that it corresponds to a Boolean algebraic assignable to genetic code and that the sub-algebra represents maximal number of independent statements defining analogs of axioms. The remaining elements would correspond to negations of these statements. I also proposed that the Boolean algebra with m=126=6× 21 bits (21 pieces consisting of 6 bits) corresponds to what I called memetic code obviously realizable as sequences of 21 DNA codons with stop codons included. Emotions and information are closely related and peptides are regarded as both information molecules and molecules of emotion.
3. This hierarchy of codes would have the additional property that the Boolean algebra at n+1:th level can be regarded as the set of statements about statements of the previous level. One would have a hierarchy representing thoughts about thoughts about.... It should be emphasized that there is no need to assume that the Hilbert's conjecture is true.

One can obtain this kind of hierarchies as hierarchies with dimensions m, 2^m, 2^{2m},... that is $n(i+1)= 2^{n(i)}$. The conditions that n(i) divides n(i+1) is non-trivial only for at the lowest step and implies that m is power of 2 so that the hierarchies starting from $m=2^k$. This is natural since Boolean algebras are involved. If n corresponds to the size scale of CD, it would come as a power of 2.

p-Adic length scale hypothesis has also led to this conjecture. A related conjecture is that the sizes of CDs correspond to secondary p-adic length scales, which indeed come as powers of two by p-adic length scale hypothesis. In case of electron this predicts that the minimal size of CD associated with electron corresponds to time scale T=.1 seconds, the fundamental time scale in living matter (10 Hz is the fundamental bio-rhythm). It seems that the basic hypothesis of TGD inspired partly by the study of elementary particle mass spectrum and basic bio-scales (there are 4 p-adic length scales defined by Gaussian Mersenne primes in the range between cell membrane thickness 10 nm and and size 2.5 µm of cell nucleus!) follow from the proposed connection between emotions and Boolean cognition.

4. NMP would be in the role of God. Strong NMP as God would force always the optimal choice maximizing negentropy gain and increasing negentropy resources of the Universe. Weak NMP as God allows free choice so that entropy gain is not be maximal and sinners populate the world. Why the omnipotent God would allow this? The reason is now obvious. Weak form of NMP makes possible the realization of Boolean algebras in terms of degrees of "feels good"! Without the God allowing the possibility to do sin there would be no emotional intelligence!

Hilbert's conjecture relates in interesting manner to space-time dimension. Suppose that Hilbert's conjecture fails and only the four lowest Mersenne integers in the hierarchy are Mersenne primes that is 3,7,127, $2^{127}-1$. In TGD one has hierarchy of dimensions associated with space-time surface coming as 0,1,2,4 plus imbedding space dimension 8. The abstraction hierarchy associated with space-time dimensions would correspond discretization of partonic 2-surfaces as point set, discretization of 3-surfaces as a set of strings connecting partonic 2-surfaces characterized by discrete parameters, discretization of space-time surfaces as a collection of string world sheet with discretized parameters, and maybe - discretization of imbedding space by a collection of space-time surfaces. Discretization means that the parameters in question are algebraic numbers in an extension of rationals associated with p-adic numbers.

In TGD framework, it is clear why imbedding space cannot be higher-dimensional and why the hierarchy does not continue. Could there be a deeper connection between these two hierarchies. For instance, could it be that higher dimensional manifolds of dimension 2×n can be represented physically only as unions of say n 2-D partonic 2-surfaces (just like 3×N dimensional space can be represented as configuration space of N point-like particles)? Also infinite primes define a hierarchy of abstractions. Could it be that one has also now similar restriction so that the hierarchy would have only finite number of levels, say four. Note that the notion of n-group and n-algebra involves an analogous abstraction hierarchy.

For details see the article Good and Evil, Life and Death. For a summary of earlier postings see Links to the latest progress in TGD.

posted by Matti Pitkanen @ 10:33 PM

4 Comments:

At 5:18 AM, Anonymous said...

Also polygonal numbers (feel good! :)) and have similar structure of statements about statements on the earlier level, e.g. triangular numbers, tetragonal etc. higher dimensional:

0   0   0   0   0   ...

1	1	1	1	1	...
1	2	3	4	5	...
1	3	6	10	15	...
1	4	10	19	31	...

etc.

The pattern can be extended also to the negative side, and there's amazing finding connecting Euler's finding about relation of "full" set of pentagonal numbers (http://en.wikipedia.org/wiki/Pentagonal_number_theorem) and sigma function(http://en.wikipedia.org/wiki/Divisor_function) together with sum of all non-cloned Egyptian fractions aka harmonic numbers and logarithmic function, resulting in elementary problem that is equivalent with Riemann hypothesis:

http://www.math.lsa.umich.edu/~lagarias/doc/elementaryrh.pdf

Nice introduction to the subject here:

https://www.youtube.com/watch?v=1mSk3J3GlA8

At 5:22 AM, Anonymous said...

Correction: "triangular, tetragonal..." -> triangular, tetrahedral...

At 6:13 AM, Anonymous said...

Scalar feel-good factor allows to feel better and better, which is good. :)

The fundamental problem with leveled empathy is that leveled negentropy between e.g. human-form emotional states lets in both feel-good and feel-bad, and raising barriers (e.g. us-against-them) against the feel-bad aspect and thus the whole holografic empathy of multi-observer negentropic field.

Empathy can thus be stated as strong form of emotional holography, and by not actively filtering out the emotions of other tribes, other species, spirits and gods, we can trust that the pain that comes in is nothing compared to the whole of love (=God). While also fully sympathizing with the fear of opening heart more and more fully also to the suffering of others, and non-judgementally allowing that basic fear the space and time and life-experience that it needs.

The Boolean aspect of what is called 'monogamy' of entangled observables seems to be the core mathematical issue relating to empathy, understood as strong holography of emotional negentropic entanglement, and monogamy of entangled observables (cf. representation theory and epistemology) is related to weak form of NMP. However, we do not need to treat strong NMP and weak NMP as either-or question, when we remember that according to Spinoza's definition, Absolute contains all attributes, and that Spinoza's Ethics is happily smiling feel-better philosophy. :)

This in mind, can we see a way to combine and relate strong NMP and weak NMP in terms of both monogamic and polyamoric negentropic entanglements of observables?

At 6:59 AM, Matpitka@luukku.com said...

To Anonymous:

At least now I am happy with weak NMP. It allows realisation of emotional Boolean intelligence rather than only the usual cold and academic one;-).

An interesting question: can one map this emotionally represented Boolean algebra to fermionic representation of Boolean algebra: m-D subspace to m-fermion state. One should pair many-fermion states representing n-bit Boolean algebra with the subspaces of space defined by projection operator. One should entangle2^n-1 many-fermion states with these 2^n-1 state functions reductions? Sounds crazy!

04/20/2015 - http://matpitka.blogspot.com/2015/04/can-one-identify-quantum-physical.html#comments

Can one identify quantum physical correlates of ethics and moral?

TGD-inspired theory of Consciousness involves a bundle of new concepts making it possible to seriously discuss quantum physical correlates of ethics and moral assuming that we live in the TGD Universe. In the following I summarize the recent understanding. I do not guarantee that I will agree with myself tomorrow since I am just going through this stuff in the updating of TGD-inspired theory of Consciousness and Quantum Biology.

Quantum ethics very briefly

Could physics generalized to a theory of consciousness allow to undersand the physical correlates of ethics and moral. The proposal is that this is the case. The basic ethical principle would be that good deeds help evolution to occur. Evolution should correspond to the increase of negentropic entanglement resources, defining negentropy sources, which I have called Akashic records.   This idea can be criticized.

1. If strong form of NMP prevails, one can worry that TGD Universe does not allow Evil at all, perhaps not even genuine free will! No-one wants Evil but Evil seems to be present in this world.
2. Could one weaken NMP so that it does not force but only allows to make a reduction to a final state characterized by density matrix which is projection operator? Self could choose whether to perform a projection to some sub-space of this subspace, say 1-D ray as in ordinary state function reduction. NMP would be like Christian God allowing the sinner to choose between Good and Evil. The final entanglement negentropy would be measure for the goodness of the deed. This is so if entanglement negentropy is a correlate for love. Deeds which are done with love would be good. Reduction of entanglement would in turn mean loneliness and separation.
3. Or could could think that the definition of good deed is as a selection between deeds, which correspond to the same maximal increase of negentropy so that NMP cannot tell what happens. For instance the density matrix operator is direct sum of projection operators of same dimension but varying coefficients and there is a selection between these. It is difficult to imagine what the criterion for a good deed could be in this case. And how self can know what is the good deed and what is the bad deed.

Good deeds would support evolution. There are many manners to interpret evolution in theTGD Universe.

1. p-Adic evolution would mean a gradual increase of the p-adic primes characterizing individual partonic 2-surfaces and therefore their size. The identification of p-adic space-time sheets as representations for cognitions gives additional concreteness to this vision. The earlier proposal that p-adic--real-phase transitions correspond to realization of intentions and formations of cognitions seems however to be wrong. Instead, adelic view that both real and p-adic sectors are present simultaneously and that fermions at string world sheets correspond to the intersection of realities and p-adicities seems more realistic.

The inclusion of phases q=exp(i2π/n) in the algebraic extension of p-adics allows to define the notion of angle in p-adic context but only with a finite resolution since only finite number of angles are represented as phases for a given value of n. The increase of the integers n could be interpreted as the emergence of higher algebraic extensions of p-adic numbers in the intersection of the real and p-adic worlds. These observations suggest that all three views about evolution are closely related.

2. The hierarchy of Planck constants suggests evolution as the gradual increase of the Planck constant characterizing p-adic space-time sheet (or partonic 2-surface for the minimal option). The original vision about this evolution was as a migration to the pages of the book like structure defined by the generalized imbedding space and has therefore quite concrete geometric meaning. It implies longer time scales of long term memory and planned action and macroscopic quantum coherence in longer scales.

   The new view is in terms of first quantum jumps to the opposite boundary of CD leading to the death of self and its re-incarnation at the opposite boundary.

3. The vision about Life as something in the intersection of real and p-adic words allows to see evolution information theoretically as the increase of number entanglement negentropy implying entanglement in increasing length scales. This option is equivalent with the second view and consistent with the first one if the effective p-adic topology characterizes the real partonic 2-surfaces in the intersection of p-adic and real worlds.

The third kind of evolution would mean also the evolution of spiritual consciousness if the proposed interpretation is correct. In each quantum jump U-process generates a superposition of states in which any sub-system can have both real and algebraic entanglement with the external world. If state function reduction process involves also the choice of the type of entanglement it could be interpreted as a choice between good and evil. The hedonistic complete freedom resulting as the entanglement entropy is reduced to zero on one hand, and the negentropic entanglement implying correlations with the external world and meaning giving up the maximal freedom on the other hand. The selfish option means separation and loneliness. The second option means expansion of consciousness - a fusion to the ocean of consciousness as described by spiritual practices.

In this framework one could understand the physics correlates of ethics and moral. The ethics is simple: evolution of consciousness to higher levels is a good thing. Anything which tends to reduce consciousness represents violence and is a bad thing. Moral rules are related to the relationship between individual and society and presumably develop via self-organization process and are by no means unique. Moral rules however tend to optimize evolution. As blind normative rules they can however become a source of violence identified as any action which reduces the level of consciousness.

There is an entire hierarchy of selves and every self has the selfish desire to survive and moral rules develop as a kind of compromise and evolve all the time. ZEO leads to the notion that I have christened cosmology of consciousness. It forces to extend the concept of society to four-dimensional society. The decisions of "me now" affect both my past and future and time like quantum entanglement makes possible conscious communication in time direction by sharing conscious experiences. One can therefore speak of genuinely four-dimensional society. Besides my next-door neighbors I had better to take into account also my nearest neighbors in past and future (the nearest ones being perhaps copies of me!). If I make wrong decisions those copies of me in future and past will suffer the most. Perhaps my personal hell and paradise are here and are created mostly by me.

What could the quantum correlates of moral be?

We make moral choices all the time. Some deeds are good, some deeds are bad. In the world of materialist there are no moral choices, the deeds are not good or bad, there are just physical events. I am not a materialist so that I cannot avoid questions such as how do the moral rules emerge and how some deeds become good and some deeds bad. Negentropic entanglement is the obvious first guess if one wants to understand emergence of moral.

1. One can start from ordinary quantum entanglement. It corresponds to a superposition of pairs of states. Second state corresponds to the internal state of the self and second state to a state of external world or biological body of self. In negentropic quantum entanglement each is replaced with a pair of sub-spaces of state spaces of self and external world. The dimension of the sub-space depends on the which pair is in question. In state function reduction one of these pairs is selected and deed is done. How to make some of these deeds good and some bad?
2. Obviously the value of $h_{eff}/h = n$ gives the criterion in the case that weak form of NMP holds true. Recall that weak form of NMP allows only the possibility to generate negentropic entanglement but does not force it. NMP is like God allowing the possibility to do good but not forcing good deeds.

Self can choose any sub-space of the subspace defined by n-dimensional projector and 1-D subspace corresponds to the standard quantum measurement. For n=1 the state function reduction leads to vanishing negentropy, and separation of self and the target of the action. Negentropy does not increase in this action and self is isolated from the target: kind of price for sin.

For the maximal dimension of this sub-space the negentropy gain is maximal. This deed is good and by the proposed criterion the negentropic entanglement corresponds to love or more generally, positively colored conscious experience. Interestingly, there are 2^n possible choices which is the dimension of Boolean algebra consisting of n independent bits. This could relate directly to fermionic oscillator operators defining basis of Boolean algebra. The deed in this sense would be a choice of how loving the attention towards system of external world is.

3. Could the moral rules of society be represented as this kind of entanglement patterns between its members? Here one of course has entire fractal hierarchy of societies corresponding different length scales. Attention and magnetic flux tubes serving as its correlates is the basic element also in TGD inspired quantum biology already at the level of bio-molecules and even elementary particles. The value of $h_{eff}/h=n$ associated with the magnetic flux tube connecting members of the pair, would serve as a measure for the ethical value of maximally good deed. Dark phases of matter would correspond to good: usually darkness is associated with bad!

4. These moral rules seem to be universal. There are however also moral rules or should one talk about rules of survival, which are based on negative emotions such as fear. Moral rules as rules of desired behavior are often tailored for the purposes of power holder. How this kind of moral rules could develop? Maybe they cannot be realized in terms of negentropic entanglement. Maybe the superposition of the allowed alternatives for the deed contains only the alternatives allowed by the power holder and the superposition in question corresponds to ordinary entanglement for which the signature is simple: the probabilities of various options are different. This forces the self to choose just one option from the options that power holder accepts. These rules do not allow the generation of loving relationship.

Moral rules seem to be generated by society, up-bringing, culture, civilization. How the moral rules develop? One can try to formulate and answer in terms of quantum physical correlates.

1. Basically the rules should be generated in the state function reductions which correspond to volitional action which corresponds to the first state function reduction to the earlier active boundary of CD. Old self dies and new self is born at the opposite boundary of CD and the arrow of time associated with CD changes.

2. The repeated sequences of state function reductions can generate negentropic entanglement during the quantum evolutions between them. This time evolution would be the analog for the time evolution defined by Hamiltonian - that is energy - associated with ordinary time translation whereas the first state function reduction at the opposite boundary inducing scaling of h_{eff} and CD would be accompanied by time evolution defined by conformal scaling generator L_0.
Note that the state at passive boundary does not change during the sequence of repeated state function reductions. These repeated reductions however change the parts of zero energy states associated with the new active boundary and generate also negentropic entanglement. As the self dies the moral choices can made if the weak form of NMP is true.

3. Who makes the moral choices? It looks of course very weird that self would apply free will only at the moment of its death or birth! The situation is saved by the fact that self has also sub-selves, which correspond to sub-CDs and represent mental images of self. We know that mental images die as also we do some day and are born again (as also we do some day) and these mental images can generate negentropic resources within CD of self.

One can argue that these mental images do not decide about whether to do maximally ethical choice at the moment of death. The decision must be made by a self at higher level. It is me who decides about the fate of my mental images - to some degree also after their death! I can choose the how negentropic the quantum entanglement characterizing the relationship of my mental image and the world outside it. I realize, that the misused idea of positive thinking seems to unavoidably creep in! I have however no intention to make money with it!

There are still many questions that are waiting for more detailed answer. These questions are also a good manner to detect logical inconsistencies.

1. What is the size of CD characterizing self? For electron it would be at least of the order of Earth size. During the lifetime of CD the size of CD increases and the order of magnitude is measured in light-life time for us. This would allow to understand our usual deeds affecting the environment in terms of our subselves and their entanglement with the external world which is actually our internal world, at least if magnetic bodies are considered.

2. Can one assume that the dynamics inside CD is independent from what happens outside CD. Can one say that the boundaries of CD define the ends of space-time or does space-time continue outside them. Do the boundaries of CD define boundaries for 4-D spotlight of attention or for one particular reality? Does the answer to this question have any relevance if everything physically testable is formulated in term physics of string world sheets associated with space-time surfaces inside CD?

   Note that the (average) size of CDs (, which could be in superposition but need not if every repeated state function reduction is followed by a localization in the moduli space of CDs) increases during the life cycle of self. This makes possible generation of negentropic entanglement between more and more distant systems. I have written about the possibility that ZEO could make possible interaction with distant civilizations (see this. The possibility of having communications in both time directions would allow to circumvent the barrier due to the finite light-velocity, and gravitational quantum coherence in cosmic scales would make possible negentropic entanglement.

3. How selves interact? CDs as spot-lights of attention should overlap in order that the interaction is possible. Formation of flux tubes makes possible quantum entanglement. The string world sheets carrying fermions also essential correlates of entanglement and the possibly entanglement is between fermions associated with partonic 2-surfaces. The string world sheets define the intersection of real and p-adic worlds, where cognition and life resides.

For details see the article Good and Evil, Life and Death.

Intentions, cognitions, time, and p-adic physics

Intentions involved time in an essential manner and this led to the idea that p-adic-to-real quantum jumps could correspond to a realization of intentions as actions. It however seems that this hypothesis posing strong additional mathematical challenges is not needed if one accepts adelic approach in which real space-time time and its p-adic variants are all present and quantum physics is adelic. I have developed the first formulation of p-adic space-time surface in and the ideas related to the adelic vision (see this, this, and this).

1. What intentions are?

One of the earlier ideas about the flow of subjective time was that it corresponds to a phase transition front representing a transformation of intentions to actions and propagating towards the geometric future quantum jump by quantum jump. The assumption about this front is un-necessary in the recent view inspired by ZEO.

Intentions should relate to active aspects of conscious experience. The question is what the quantum physical correlates of intentions are and what happens in the transformation of intention to action.

1. The old proposal that p-adic-to-real transition could correspond to the realization of intention as action. One can even consider the possibility that the sequence of state function reductions decomposes to pairs real-to-padic and p-adic-to-real transitons. This picture does not explain why and how intention gradually evolves stronger and stronger, and is finally realized. The identification of p-adic space-time sheets as correlates of cognition is however natural.
2. The newer proposal, which might be called adelic, is that real and p-adic space-time sheets form a larger sensory-cognitive structure: cognitive and sensory aspects would be simultaneously present. Real and p-adic space-time surfaces would form single coherent whole which could be called adelic space-time. All p-adic manifolds could be present and define kind of chart maps about real preferred extremals so that they would not be independent entities as for the first option. The first objection is that the assignment of fermions separately to the every factor of adelic space-time does not make sense. This objection is circumvented if fermions belong to the intersection of realities and p-adicities.

This makes sense if string world sheets carrying the induced spinor fields define seats of cognitive representations in the intersection of reality and p-adicities.

Cognition would be still associated with the p-adic space-time sheets and sensory experience with real ones. What can sensed and cognized would reside in the intersection.

Intention would be however something different for the adelic option. The intention to perform quantum jump at the opposite boundary would develop during the sequence of state function reductions at fixed boundary and eventually NMP would force the transformation of intention to action as first state function reduction at opposite boundary. NMP would guarantee that the urge to do something develops so strong that eventually something is done.

Intention involves two aspects. The plan for achieving something which corresponds to cognition and the will to achieve something which corresponds to emotional state. These aspects could correspond to p-adic and real aspects of intentionality.

2. p-Adic physics as physics of only cognition?

There are two views about p-adic-real correspondence corresponding to two views about p-adic physics. According to the first view p-adic physics defines correlates for both cognition and intentionality whereas second view states that it provides correlates for cognition only.

1. Option A: The older view is that p-adic -to-real transitions realize intentions as actions and opposite transitions generate cognitive representations. Quantum state would be either real or p-adic. This option raises hard mathematical challenges since scattering amplitudes between different number fields are needed and the needed mathematics might not exist at all.
2. Option B: Second view is that cognition and sensory aspects of experience are simultaneously present at all levels and means that real space-time surface and their real counterparts form a larger structure in the spirit of what might be called Adelic TGD. p-Adic space-time charts could be present for all primes. It is of course necessary to understand why it is possible to assign definite prime to a given elementary particle.

This option could be developed by generalizing the existing mathematics of adeles by replacing number in given number field with a space-time surface in the imbedding space corresponding that number field. Therefore this option looks more promising. For this option also the development of intention can be also understood. The condition that the scattering amplitudes are in the intersection of reality and p-adicities is very powerful condition on the scattering amplitudes and would reduce the realization of number theoretical universality and p-adicization to that for string world sheets and partonic 2-surfaces.

For instance, the difficult problem of defining p-adic analogs of topological invariant would trivialize since these invariants (say genus) have algebraic representation for 2-D geometries. 2-dimensionality of cognitive representation would be perhaps basically due to the close correspondence between algebra and topology in dimension D=2.

Most of the following considerations apply in both cases.

3. Some questions to ponder

The following questions are part of the list of question that one must ponder.

a) Do cognitive representations reside in the intersection of reality and p-adicities?

The idea that cognitive representation reside in the intersection of reality and various p-adicities is one of the key ideas of TGD-inspired theory of Consciousness.

1. All quantum states have vanishing total quantum numbers in ZEO, which now forms the basis of quantum TGD (see this). In principle conservation laws do not pose any constraints on possibly occurring real--p-adic transitions (Option A) if they occur between zero energy states.

   On the other hand, there are good hopes about the definition of p-adic variants of conserved quantities by algebraic continuation since the stringy quantal Noether charges make sense in all number fields if string world sheets are in the real--p-adic intersection. This continuation is indeed needed if quantum states have adelic structure (Option B). In accordance with this quantum classical correspondence (QCC) demands that the classical conserved quantities in the Cartan algebra of symmetries are equal to the eigenvalues of the quantal charges.

2. The starting point is the interpretation of fermions as correlates for Boolean cognition and p-adic space-time sheets space-time correlates for cognitions (see this). Induced spinor fields are localized at string world sheets, which suggests that string world sheets and partonic 2-surfaces define cognitive representations in the intersection of realities and p-adicities. The space-time adele would have a book-like structure with the back of the book defined by string world sheets.

3. At the level of partonic 2-surfaces common rational points (or more generally common points in algebraic extension of rationals) correspond to the real--p-adic intersection. It is natural to identify the set of these points as the intersection of

string world sheets and partonic 2-surfaces at the boundaries of CDs. These points would also correspond to the ends of strings connecting partonic 2-surfaces and the ends of fermion lines at the orbits of partonic 2-surfaces (at these surfaces the signature of the induced 4-metric changes). This would give a direct connection with fermions and Boolean cognition.

1. For option A the interpretation is simple. The larger the number of points is, the higher the probability for the transitions to occur. This because the transition amplitude must involve the sum of amplitudes determined by data from the common points.
2. For option B the number of common points measures the goodness of the particular cognitive representation but does not tell anything about the probability of any quantum transition. It however allows to discriminate between different p-adic primes using the precision of the cognitive representation as a criterion. For instance, the non-determinism of Kähler action could resemble p-adic non-determinism for some algebraic extension of p-adic number field for some value of p. Also the entanglement assignable to density matrix which is n-dimensional projector would be negentropic only if the p-adic prime defining the number theoretic entropy is divisor of n. Therefore also entangled quantum state would give a strong suggestion about the value of the optimal p-adic cognitive representation as that associated with the largest power of p appearing in n.

b) Could cognitive resolution fix the measurement resolution?

For p-adic numbers the algebraic extension used (roots of unity fix the resolution in angle degrees of freredom and pinary cutoffs fix the resolution in "radial" variables which are naturally positive. Could the character of quantum state or perhaps quantum transition fix measurement resolution uniquely?

1. If transitions (state function reductions) can occur only between different number fields (Option A), discretization is un-avoidable and unique if maximal. For real-real transitions the discretization would be motivated only by finite measurement resolution and need be neither necessary nor unique. Discretization is required and unique also if one requires adelic structure for the state space (Option B). Therefore both options A and B are allowed by this criterion.
2. For both options cognition and intention (if p-adic) would be one half of existence and sensory perception and motor actions would be second half of existence at fundamental level. The first half would correspond to sensory experience and motor action as time reversals of each other. This would be true even at the level of elementary particles, which would explain the amazing success of p-adic mass calculations.
3. For option A the state function reduction sequence would correspond to a formation of p-adic maps about real maps and real maps about p-adic maps: real $\rightarrow$ p-adic $\rightarrow$ real $\rightarrow$..... For option B it would correspond the sequence adelic $\rightarrow$ adelic $\rightarrow$ adelic $\rightarrow$.....
4. For both options p-adic and real physics would be unified to single coherent whole at the fundamental level but the adelic option would be much simpler. This kind of unification is highly suggestive - consider only the success of p-adic mass calculations - but I have not really seriously considered what it could mean.

c) What selects the preferred p-adic prime?

What determines the p-adic prime or preferred p-adic prime assignable to the system considered? Is it unique? Can it change?

1. An attractive hypothesis is that the most favorable p-adic prime is a factor of the integer n defining the dimension of the n× n density matrix associated with the flux tubes/fermionic strings connecting partonic 2-surfaces: the presence of fermionic strings already implies at least two partonic 2-surfaces. During the sequence of reductions at same boundary of CD n receives additional factors so that p cannot change. If wormhole contacts behave as magnetic monopoles there must be at least two of them connected by monopole flux tubes. This would give a connection with negentropic entanglement and for $h_{eff}/h=n$ to quantum criticality, dark matterm and hierarchy of inclusions of HFFs.
2. Second possibility is that the classical non-determinism making itself visible via super-symplectic invariance acting as broken conformal gauge invariance has same character as p-adic non-determinism for some value of p-adic prime. This would mean that p-adic space-time surfaces would be especially good representations of real space-time sheets. At the lowest level of hierarchy this would mean large number of common points. At higher levels large number of common parameter values in the algebraic extension of rationals in question.

d) How finite measurement resolution relates to hyper-finite factors?

The connection with hyper-finite factors suggests itself.

1. Negentropic entanglement can be said to be stabilized by finite cognitive resolution if hyper-finite factors are associated with the hierarchy of Planck constants and cognitive resolutions. For HFFs the projection to single ray of state space in state function reduction is replaced with a projection to an infinite-dimensional sub-space whose von Neumann dimension is not larger than one.
2. This raises interesting question. Could infinite integers constructible from infinite primes correspond to these infinite dimensions so that prime p would appear as a factor of this kind of infinite integer? One can say that for inclusions of hyperfinite factors the ratio of dimensions for including and included factors is quantum dimension which is algebraic number expressible in terms of quantum phase $q=\exp(i2\pi/n)$. Could n correspond to the integer ratio $n=n_f/n_i$ for the integers characterizing the sub-algebra of super-symplectic algebra acting as gauge transformations?

4. Generalizing the notion of p-adic space-time surface

The notion of p-adic manifold \citealb/picosahedron is an attempt to formulate p-adic space-time surfaces identified as preferred extremal of p-adic variants of p-adic field equations as cognitive charts of real space-time sheets. Here the essential point is that p-adic variants of field equations make sense: this is due to the fact that induced metric and induced gauge fields make sense (differential geometry exists p-adically unlike global geometry involving notions of lengths, area, etc does not exist: in particular the notion of angle and conformal invariance make sense).

The second key element is finite resolution so that p-adic chart map is not unique. Same applies to the real counterpart of p-adic extremal and having representation as space-time correlate for an intention realized as action.

The discretization of the entire space-time surface proposed in the formulation of p-adic manifold concept (see this) looks too naive an approach. It is plausible that one has an abstraction hierarchy for discretizations at various abstraction levels.

1. The simplest discretization would occur at space-time level only at partonic 2-surfaces in terms of string ends identified as algebraic points in the extension of p-adics used. For the boundaries of string world sheets at the orbits of partonic 2-surface one would have discretization for the parameters defining the boundary curve. By field equations this curve is actually a segment of light-like geodesic line and characterized by initial light-like 8-velocity, which should be therefore a number in algebraic extension of rationals. The string world sheets should have similar parameterization in terms of algebraic numbers.
By conformal invariance the finite-dimensional conformal moduli spaces and topological invariants would characterize string world sheets and partonic 2-surfaces. The p-adic variant of Teichmueller parameters was indeed introduced in p-adic mass calculations and corresponds to the dominating contribution to the particle mass (see this and this).

2. What might be called co-dimension 2 rule for discretization suggests itself. Partonic 2-surface would be replaced with the ends of fermion lines at it or equivalently: with the ends of space-like strings connecting partonic 2-surfaces at it. 3-D partonic orbit would be replaced with the fermion lines at it. 4-D space-time surface would be replaced with 2-D string world sheets. Number theoretically this would mean that one has always commutative tangent space. Physically the condition that em charge is well-defined for the spinor modes would demand co-dimension 2 rule.

3. This rule would reduce the real-p-adic correspondence at space-time level to construction of real and p-adic space-time surfaces as pairs to that for string world sheets and partonic 2-surfaces determining algebraically the corresponding space-time surfaces as preferred extremals of Kähler action. Strong form of holography indeed leads to the vision that these geometric objects can be extended to 4-D space-time surface representing preferred extremals.

4. In accordance with the generalization of AdS/CFT correspondence to TGD framework cognitive representations for physics would involve only partonic 2-surfaces and string world sheets. This would tell more about cognition rather than Universe. The 2-D objects in question would be in the intersection of reality and p-adicities and define cognitive representations of 4-D physics. Both classical and quantum physics would be adelic.

5. Space-time surfaces would not be unique but possess a degeneracy corresponding to a sub-algebra of the super-symplectic algebra isomorphic to it and and acting as conformal gauge symmetries giving rise to n conformal gauge invariance classes. The conformal weights for the sub-algebra would be n-multiples of those for the entire algebra and n would correspond to the effective Planck constant $h_{eff}/h=n$. The

Clearly, very many big ideas behind TGD and TGD inspired theory of consciousness would have this picture as a Boolean intersection.